Study Guide

for use wi

The Physical Universe

Eleventh Edition

Konrad B. Krauskopf

Arthur Beiser

Prepared by
Steven D. Carey
University of Mobile

Mc Graw Hill **Higher Education**

Boston Burr Ridge, IL Dubuque, IA Madison, WI New York San Francisco St. Louis
Bangkok Bogotá Caracas Kuala Lumpur Lisbon London Madrid Mexico City
Milan Montreal New Delhi Santiago Seoul Singapore Sydney Taipei Toronto

The McGraw·Hill Companies

Study Guide for use with
THE PHYSICAL UNIVERSE, ELEVENTH EDITION
Konrad B. Krauskopf and Arthur Beiser

Published by McGraw-Hill Higher Education, an imprint of The McGraw-Hill Companies, Inc., 1221 Avenue of the Americas, New York, NY 10020.

RECYCLED This book is printed on recycled, acid-free paper containing 10% postconsumer waste.

1 2 3 4 5 6 7 8 9 0 QPD/QPD 0 9 8 7 6 5 4

ISBN 0-07-297993-3

www.mhhe.com

CONTENTS

PREFACE

This study guide accompanies *The Physical Universe*, eleventh edition, by Krauskopf and Beiser. Each of the study guide's chapters is keyed to its counterpart in the text and contains a chapter summary, a sentence outline of the chapter, and additional review questions and explanatory materials designed to enhance the student's understanding of the key terms and main concepts presented in *The Physical Universe*. All chapters contain links to relevant internet sites, many of which are interactive and give the student a "hands-on" feel for the material.

This study guide differs from most others in several respects. Key terms, instead of appearing as a simple list are defined in the chapter outline and are identified by boldface type. Representative example problems are solved in a step-by-step fashion so that students with limited backgrounds in mathematics can enhance their problem-solving skills by seeing how the solutions are derived. A special section containing inexpensive and easy-to-do physical science experiments has been added for those students majoring in elementary or early childhood education.

Many people assisted in the preparation of this edition of the study guide. Arthur Beiser reviewed the manuscript and offered valuable criticisms and suggestions. Mary Hurley of McGraw-Hill Higher Eduction provided much needed encouragement and support. Maura Ives proofread portions of the manuscript and offered grammatical advice. To these individuals, anonymous reviewers, colleagues, and family, I am sincerely grateful.

Steven D. Carey
Department of Natural Sciences
University of Mobile

Chapter 1
THE SCIENTIFIC METHOD

OUTLINE

How Scientists Study Nature
1.1 The Scientific Method
1.2 Why Science Is Successful

The Solar System
1.3 A Survey of the Sky
1.4 The Ptolemaic System
1.5 The Copernican System
1.6 Kepler's laws
1.7 Why Copernicus Was Right

Universal Gravitation
1.8 What is Gravity?
1.9 Why the Earth Is Round
1.10 The Tides
1.11 The Discovery of Neptune
1.12 The SI System

GOALS

When you have completed Chapter One you should be able to:

1. Outline the scientific method.
2. Explain why the scientific method has been more successful than other approaches to understanding the universe.
3. Distinguish between a law and a theory.
4. Discuss the role of a model in formulating a scientific theory.
5. Give the reason why Polaris is the heavenly body that remains most nearly stationary in the sky.
6. Define constellation.
7. Tell how to distinguish planets from stars by observations of the night sky made several weeks or months apart.
8. Compare how the ptolemaic and copernican systems account for the observed motions of the sun, moon, planets, and stars across the sky.
9. Explain the significance of Kepler's laws.
10. State why the copernican system is considered correct.
11. Define day and year.
12. Define fundamental force.
13. Explain why the earth is round but not a perfect sphere.
14. Explain the origin of tides.
15. Explain in terms of the scientific method why the discovery of Neptune was so important in confirming the law of gravity.
16. Change the units in which a quantity is expressed from those of one system of units to those of another system.

CHAPTER SUMMARY

This chapter introduces the **scientific method**, the process scientists use to interpret the physical universe. Science is a living body of knowledge whose **laws** and **theories** are subject to constant test and change. Although science can never arrive at an "ultimate truth," it has nevertheless successfully explained the natural world and has improved the quality of life. The scientific method is illustrated by the acceptance of the **copernican model** of the universe over the ancient Greek **ptolemaic model**. In turn, this advance led to Kepler's discovery of the **laws of planetary motion** and Newton's discovery of the **law of gravity**. Gravity is one of four **fundamental forces**. The law of gravity has been verified by accounting for the occurrence of the tides, the shape of the earth, and by its successful application in the discovery of the planet Neptune. The International System (SI) of **units** is introduced, and examples of SI units are given.

CHAPTER OUTLINE

1-1. The Scientific Method

A. The **scientific method** consists of four steps:
   1. **Formulation of a problem**
   2. **Observation and experiment**
   3. **Interpretation**
   4. **Testing the interpretation**

B. An initial scientific interpretation is called the **hypothesis**.

C. A **law** describes a relationship or regularity between naturally occurring phenomena.

D. A **theory** explains why a phenomenon or a set of phenomena occurs.

E. Scientists often use **models** to simplify complex situations.

F. Newton chose an oval called an **ellipse** as a model of the earth's orbit.

1-2. Why Science Is Successful

A. A scientific law or theory, if refuted by contrary evidence, must be modified or discarded.

B. The work of scientists is open to review, test, and change.

C. Science has provided an understanding of the natural world and a sophisticated technology.

D. Scientific laws and theories are not accepted as "absolute truth" and therefore differ from belief systems.

1-3. A Survey of the Sky

A. To an observer north of the equator, the position of the North Star, or **Polaris**, changes very little, and the whole nighttime sky appears to revolve around Polaris.

B. The **constellations** are easily recognized groups of stars and are useful as labels for regions of the sky.
C. The **planets** visible to the naked eye (Mercury, Venus, Mars, Jupiter, and Saturn) appear to drift in a generally eastward motion relative to the stars; however, each planet at times appears to head westward briefly, and its path across the sky resembles a series of loops.

1-4. The Ptolemaic System

A. Ptolemy of Alexandria (2nd century A.D.) described the universe in the *Almagest*.
   1. The earth is the center of the universe.
   2. The sun, stars, and planets revolve around the earth.
   3. The orbits of the planets are circular.
B. According to Ptolemy, the planets as they orbit the earth travel in a series of loops (epicycles).
C. The **ptolemaic system** had the components of a valid theory:
   1. It was based on observation.
   2. It apparently accounted for known celestial motions.
   3. It made predictions that could be tested.

1-5. The Copernican System

A. The ptolemaic system failed to make accurate predictions of planetary positions.
B. Nicholaus Copernicus (1473-1543) developed a new theory of the universe:
   1. The earth and planets follow circular orbits around the sun.
   2. The earth rotates on its axis once every 24 hours.
   3. The earth's rotation explains the daily rising and setting of celestial bodies.
   4. Irregular movements of the planets are a result of the combination of their motions around the sun and the change in position of the earth in its orbit.
C. The **copernican system** was attacked by religious leaders and by other supporters of the ptolemaic system.

1-6. Kepler's Laws

A. Johannes Kepler (1571-1630), using Tycho Brahe's improved measurements of planetary motion, found fault with the copernican system.
B. Kepler's calculations resulted in the discovery of three laws of planetary motion:
   1. The paths of the planets around the sun are ellipses with the sun at one focus.
   2. A planet moves so that its radius vector sweeps out equal areas in equal times.
   3. The ratio between the square of the time needed by a planet to make a revolution around the sun and the cube of its average distance from the sun is the same for all the planets.

C. Kepler's laws agreed with past observations of planetary positions and made accurate predictions of future planetary movements.

1-7. Why Copernicus Was Right

A. There is direct evidence that the earth rotates and revolves around the sun.
B. There is direct evidence for the motions of the moon and the other planets.

1-8. What Is Gravity?

A. Isaac Newton's (1642-1727) discovery of the law of gravity was dependent upon Copernicus's model of the solar system.
B. Gravity is a **fundamental force**.
    1. A fundamental force cannot be explained in terms of any other force.
    2. There are four fundamental forces:
        a. Gravitational
        b. Electromagnetic
        c. Weak
        d. Strong
C. Gravity is thought to be a universal force because:
    1. Observed star systems and galaxies behave as if influenced by gravity.
    2. Matter appears to be the same throughout the universe; therefore, gravitational attraction must also be universal.
    3. There is no evidence that gravity is not universal.

1-9. Why the Earth Is Round

A. The theory of gravity accounts for the earth's shape; the earth is round because gravity squeezes it into a spherical shape.
B. The earth is not a perfect sphere because its spinning motion causes it to bulge slightly at the equator and flatten slightly at the poles.

1-10. The Tides

A. The law of gravity successfully explains the occurrence of the tides.
B. The earth's tides are the result of the gravitational attraction of the moon and the sun.
C. Coastal areas experience two high tides and two low tides each day.
D. The relative positions of the earth, sun, and moon produce different tides.
    1. Unusually high (and low) **spring tides** occur bimonthly when the moon and sun are aligned with the earth.
    2. Weak **neap tides** occur bimonthly when the sun and moon pull at right angles to each other in regard to earth.

1-11. The Discovery of Neptune

A. Discrepancies in the predicted orbit of Uranus led to two hypotheses:
   1. The law of gravity is wrong.
   2. An unknown body is exerting a gravitational pull on Uranus.

B. Calculations based on the law of gravity predicted the position of an unknown body.

C. The prediction was tested, resulting in the discovery of Neptune.

1-12. The SI System

A. A measurement consists of a number and a **unit** or standard quantity.

B. Scientists and almost all nations use the SI or International System of measurement.
   1. Examples of SI units include:
      a. **Meter** (m) for length
      b. **Second** (s) for time
      c. **Kilogram** (kg) for mass
      d. **Joule** (J) for energy
      e. **Watt** (W) for power
   2. Units in the SI are based on subdivisions and multiples of 10.

KEY TERMS AND CONCEPTS

The questions in this section will help you review the key terms and concepts from Chapter 1.

Short Answer

Each statement below represents a specific step in the scientific method. Decide which one of the four steps—**formulating a problem**, **observation and experiment**, **interpretation**, or **testing the interpretation**—applies to the statement and write the name of the step in the space below the statement.

1. Isaac Newton wonders if the force holding the planets in orbit around the sun is the same force that pulls objects to the earth's surface.

2. Johannes Kepler calculates planetary motion from the data of Tycho Brahe.

3. Albert Einstein's prediction that light is affected by gravity is verified by experiment.

4. Ptolemy of Alexandria describes a model of the universe with the earth stationary at its center and with the sun revolving around the earth.

5. A scientist asks what the relationship is between the moon and the earth's tides.

6. Tycho Brahe's observatory was able to determine celestial angles to better than 1/100 of a degree.

7. Urbain Leverrier of France and John Couch Adams of England propose that an unseen and unknown celestial body is responsible for the observed discrepancies in the orbit of Uranus and calculate the unknown body's position.

8. Johann Gottfried Galle, using the prediction made by Leverrier, discovers the planet Neptune.

9. Nicholaus Copernicus asks if the planets move around the sun rather than around the earth.

10. Copernicus proposes that the earth rotates once on its axis every 24 hours.

Multiple Choice

Circle the best answer for each of the following questions.

1. Which statement best characterizes the nature of science?
   a. the laws and theories of science are based on belief and speculation
   b. science is a living body of knowledge, not a set of unchanging ideas
   c. science has done a poor job of explaining physical phenomena and has failed to improve the quality of human life
   d. science is superior to other aspects of human culture such as religion, art, and music

2. A scientific law
   a. is seldom based upon experimental evidence since it can never be considered to be absolutely true
   b. usually states a regularity or relationship that describes how nature behaves in a certain specific way
   c. explains why certain phenomena in nature take place
   d. is known to be true beyond a shadow of a doubt

3. The ptolemaic system fulfilled the requirements of a scientific theory because
   a. its explanations of celestial motions, based on observations, resulted in testable predictions
   b. it was presented openly for public inspection when included in Ptolemy's *Almagest*
   c. it was believed to represent an accurate view of the universe and solar system by religious leaders and learned scholars of the time
   d. it provided a "common sense" explanation concerning the observed motions of heavenly bodies

4. The principle known as Occam's razor states that
   a. the most complicated scientific explanation for a given phenomenon is likely to be correct
   b. a scientific hypothesis that makes common sense is most likely to be correct
   c. scientific inquiry can never lead to a complete understanding of the natural world because it is impossible to precisely measure any physical parameter
   d. the simplest scientific explanation for a phenomenon is most likely to be correct

5. Newton's discovery of the law of gravity was dependent upon
   a. the discovery of the planet Neptune
   b. an understanding of the shape of the earth
   c. the development of the copernican system of the solar system
   d. the law that states what goes up must come down

6. The time it takes a planet to make one complete trip around the sun is called the planet's
   a. revolution
   b. orbit
   c. rotation
   d. period

7. The modern version of the copernican system is considered to be correct because
   a. most people believe that the copernican system is correct, and the majority viewpoint rules
   b. predictions of planetary motions based on the copernican system proved to be correct
   c. there is direct evidence that the earth rotates and the planets revolve around the sun
   d. a committee of scientists has certified that the copernican system is correct

8. When observable evidence does not agree with a scientific theory
   a. the truthfulness of the evidence must be questioned since a theory is never wrong
   b. the evidence is ignored and the theory remains valid
   c. the theory is reduced in status to a hypothesis
   d. the theory must be modified or discarded

9. Spring tides
   a. occur only in the spring
   b. occur when the sun and moon are in line with the earth
   c. have a low range between high and low water
   d. occur when the sun and moon are 90° apart relative to the earth

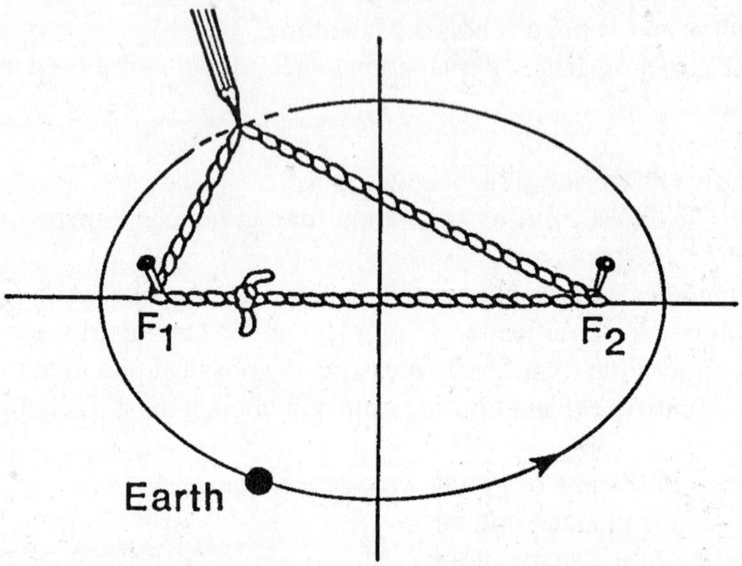

Refer to the above drawing to answer multiple choice questions 10-12.

10. The points corresponding to the positions of the tacks are called
    a. epicycles
    b. focuses
    c. planets
    d. orbits

11. The sun would occupy a position
    a. at either point F_1 or F_2
    b. at the center of the ellipse
    c. at the pencil
    d. somewhere on the circumference of the ellipse

12. The drawing is a visual representation of
    a. Newton's law of gravity
    b. Kepler's first law of planetary motion
    c. Kepler's second law of planetary motion
    d. Kepler's third law of planetary motion

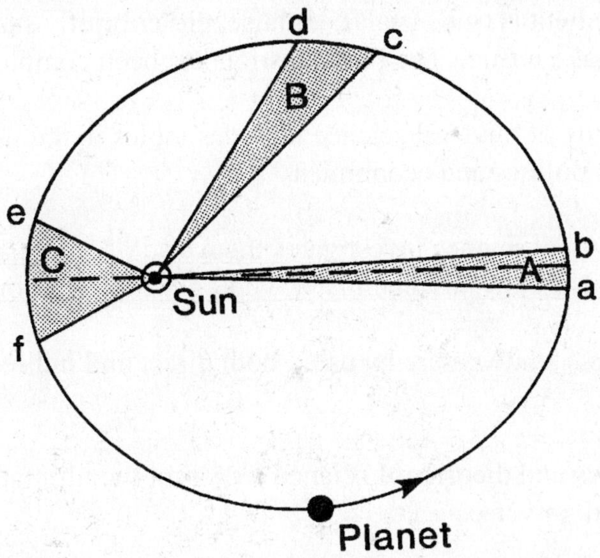

Refer to the above drawing to answer multiple choice questions 13-15. Let the amount of time it takes the planet to travel the distance from a to b, from c to d, and from e to f to be the same.

13. Section C is
    a. greater in area than section A
    b. greater in area than section B
    c. lesser in area than either section A or B
    d. the same in area as section A or B

14. The planet would be traveling the fastest along that portion of its orbit represented by the distance between the letters
    a. a and b
    b. c and d
    c. e and f
    d. none of the answers is correct because the planet's speed remains constant along its orbit

15. If the average orbit radius of the planet is increased, the period of the planet will
    a. decrease
    b. increase
    c. remain unchanged
    d. decrease or increase depending on the mass of the planet

True or False

Decide whether each statement is true or false. If false, either briefly state why it is false or correct the statement to make it true. The first question has been completed as an example.

FALSE 1. The study of physical science includes topics in the world of human events such as politics and economics.

Physical science investigates the world of nature and covers topics in the fields of physics, chemistry, geology, and astronomy.

_____ 2. Scientists study nature by using both direct and indirect methods.

_____ 3. The laws and theories of science represent the ultimate truth about our physical universe.

_____ 4. According to Kepler's third law, the speed that a planet travels in its orbit around the sun depends upon the planet's mass.

_____ 5. Kepler's laws of planetary motion are explained by Newton's discovery of the law of gravity.

_____ 6. When first proposed, a scientific interpretation is usually called a theory.

_____ 7. Astrology has been used to make consistently accurate predictions about the future.

_____ 8. The SI unit for power is the joule.

_____ 9. The great advantage of SI units is that their subdivisions and multiples are in steps of 10.

_____ 10. The SI unit for time is the minute.

Fill in the Blank

1. An imaginary line joining a planet with the sun is known as its ____________________ vector.

2. ____________________ are groups of stars named after objects, animals, people, or mythological beings.

3. ____________________, or the North Star, appears to move very little while the whole nighttime sky appears to revolve around it.

4. In the copernican system, the orbits of the planets are ____________________ in shape.

5. A ____________________ is any circle on the earth's surface whose center is the earth's center.

6. The ____________________ is a great circle midway between the north and south poles.

7. A ____________________ is a great circle that passes through both the north and south poles.

8. The ____________________ of a point on the earth's surface is the angular distance between a meridian through this point and the prime meridian.

9. The ____________________ of a place on the earth's surface is the angular distance of the place north or south of the equator.

10. The ____________________ is a unit of speed equal to 1 nautical mile per hour.

Matching

Match the name of the person on the left with the description on the right.

1._____ Johann Gottfried Galle

2._____ Tycho Brahe

3._____ Isaac Newton

4._____ Ptolemy of Alexandria

5._____ Johannes Kepler

6._____ Nicolaus Copernicus

a. believed the earth was stationary at the center of the universe
b. discovered the law of gravity
c. discovered the planet Neptune
d. proposed that the earth rotates on its axis and, along with the other planets, revolves around the sun
e. Danish astronomer whose measurements of planetary motion resulted in revision of the copernican system
f. discovered the three laws of planetary motion

WEB LINKS

See interactive demonstrations of Kepler's laws at these web sites

http://www.physics.nwu.edu/ugrad/vpl/mechanics/planets.html

http://www.physics.sjsu.edu/tomley/kepler.html

http://www.phy.hw.ac.uk/~peckham/demos/orbit_simulator/orbit1.html

Online guides to the SI are available at:

http://www.cofc.edu/~frysingj/SIguidelines.htm

http://www.alcyone.com/max/reference/physics/units.html

ANSWER KEY

Short Answer

1. Formulating a Problem
2. Experiment and Observation
3. Testing the Interpretation
4. Interpretation
5. Formulating a Problem
6. Experiment and Observation
7. Interpretation
8. Testing the Interpretation
9. Formulating a Problem
10. Interpretation

Multiple Choice

1. b 2. b 3. a 4. d 5. c 6. d 7. c 8. d 9. b 10. b 11. a 12. b 13. d 14. c 15. b

True or False

1. False. See explanatory notes given in the example.
2. True
3. False. There are no ultimate truths in science; scientific laws and theories are valid only as long as no contrary evidence comes to light.
4. False. The speed that a planet travels in its orbit around the sun depends upon its distance from the sun.
5. True
6. False. When first proposed, a scientific interpretation is usually called a hypothesis.
7. False. There is no evidence that astrology can predict the future.
8. False. The SI unit for power is the watt.
9. True.
10. False. The SI unit for time is the second.

Fill in the Blank

1. radius
2. constellations
3. Polaris
4. circular
5. great circle
6. equator
7. meridian
8. longitude
9. latitude
10. knot

Matching

1. c 2. e 3. b 4. a 5. f 6. d

Chapter 2
MOTION

OUTLINE

Describing Motion
2.1 Speed
2.2 Vectors
2.3 Acceleration
2.4 Distance, Time, and Acceleration

Acceleration of Gravity
2.5 Free Fall
2.6 Air Resistence

Force and Motion
2.7 First law of Motion
2.8 Mass
2.9 Second law of Motion
2.10 Mass and Weight
2.11 Third law of Motion

Gravitation
2.12 Circular Motion
2.13 Newton's Law of Gravity
2.14 Artificial Satellites

GOALS

1. Distinguish between instantaneous and average speeds.
2. Use the formula $v = d/t$ to solve problems that involve distance, time, and speed.
3. Distinguish between scalar and vector quantities and give several examples of each.
4. Define acceleration and find the acceleration of an object whose speed is changing.
5. Explain what is meant by the acceleration of gravity.
6. Describe the effect of air resistance on falling objects.
7. Define force and indicate its relationship to the first law of motion.
8. Discuss the significance of the second law of motion, $F = ma$.
9. Distinguish between mass and weight and be able to find the weight of an object of given mass.
10. Use the third law of motion to relate action and reaction forces.
11. Explain the significance of centripetal force in motion along a curved path.
12. Relate the centripetal force on an object moving in a circle to its mass, speed, and the radius of the circle.
13. State Newton's law of gravity and describe how gravitational forces vary with distance.
14. Account for the ability of a satellite to orbit the earth without either falling to the ground or flying off into space.
15. Define escape speed.

CHAPTER SUMMARY

This chapter discusses the concept of motion, both straight-line and circular. Motion is described in terms of **speed**, **velocity**, and **acceleration**. **Scalar quantities** are differentiated from **vector quantities**, and the use of **vectors** to represent vector quantities is presented. The downward **acceleration of gravity** is shown to influence the motion of objects in free fall near the earth's surface. The behavior of moving bodies is summarized by Newton's three **laws of motion**. The concept of **inertia** is developed in the **first law of motion**, and the relationship between **force, mass,** and **acceleration** is operationally defined by the **second law of motion**. **Weight** is defined as a force, and the relationship between weight and mass is described operationally in terms of the second law. The concept that forces come in pairs consisting of an **action force** and a **reaction force** is developed in the **third law of motion**. Circular motion is shown to be the result of **centripetal force** acting upon a moving body.

CHAPTER OUTLINE

2-1. Speed

A. The **speed** of a moving object is the rate at which it covers distance:

$$v = \frac{d}{t}$$

where v = speed, d = distance, and t = time.

B. **Average speed** is the total distance traveled by an object divided by the time taken to travel that distance.

C. **Instantaneous speed** is an object's speed at a given instant of time.

2-2. Vectors

A. The **magnitude** of a quantity tells how large the quantity is.

B. There are two types of quantities:
   1. **Scalar quantities** have magnitude only
   2. **Vector quantities** have both magnitude and direction

C. **Velocity** is a vector quantity that includes both speed and direction.

D. A **vector** can be represented by an arrowhead line whose length is proportional to the magnitude of some quantity and whose direction is that of the quantity.

2-3. Acceleration

A. The **acceleration** of an object is the rate of change of its velocity and is a vector quantity.

B. For straight-line motion, average acceleration is the rate of change of speed:

$$a = \frac{v_f - v_i}{t}$$

where a = acceleration, t = time, v_f = final speed, and v_i = initial speed.

2-4. Distance, Time, and Acceleration

A. Speed and acceleration are defined quantities.
B. Relating speed and acceleration to each other and to time (a measurable quantity) allows the formation of equations that can answer certain questions involving distance, time, and acceleration.

2-5. Free Fall

A. The **acceleration of gravity** (g) for objects in free fall at the earth's surface is 9.8 m/s^2.
B. An object thrown horizontally has a vertical acceleration of 9.8 m/s^2 and a constant horizontal velocity; its trajectory is a curved path.
C. An object thrown upward falls back to earth at an acceleration of 9.8 m/s^2 and returns to its starting point with the same speed at which it was thrown.
D. An object thrown downward has a final speed that is the sum of its original speed plus the increase in speed due to the acceleration of gravity.
E. An object thrown upward at an angle to the ground follows a curved path called a **parabola.**

2-6. Air Resistance

A. Air resistance prevents a falling object from reaching the full speed produced by the acceleration of gravity.
B. Air resistance increases with the speed of a moving object.
C. **Terminal speed** is the maximum speed a falling object obtains when the force due to downward acceleration of gravity is balanced by the upward force of air resistance.
D. The terminal speed for any object depends on the object's size, shape, and mass.

2-7. First Law of Motion

A. Isaac Newton formulated the three **laws of motion.**
B. The **first law of motion** states: If no net force acts on it, an object at rest remains at rest and an object in motion remains in motion at a constant velocity.
C. A **force** is any influence that can cause an object to be accelerated.
D. Every acceleration can be traced to the action of a force.

2-8. Mass

A. **Inertia** is the apparent resistance an object offers to any change in its state of rest or motion.

B. The **mass** of a body is the property of matter that manifests itself as inertia.
   1. Mass may be thought of as a quantity of matter.
   2. The greater an object's mass, the greater the object's inertia.

C. The SI unit for mass is the kilogram (kg).

2-9. Second Law of Motion

A. Newton's **second law of motion** states: The net force on an object equals the product of the mass and the acceleration of the object. The direction of the force is the same as that of the acceleration.

B. The second law of motion can be expressed in either of these ways:

$$a = \frac{F}{m}$$

where a = acceleration, F = force, and m = mass, or

$$F = ma$$

where F = force, m = mass, and a = acceleration.

C. The SI unit of force is the **newton** (N):

$$1 \text{ newton} = 1 \text{ N} = 1 \text{ (kg)(m/s}^2)$$

D. The **pound** (lb) is the unit of force in the British system of measurement:

$$1 \text{ lb} = 4.45 \text{ N} \quad (1 \text{ N} = 0.225 \text{ lb})$$

2-10. Mass and Weight

A. The **weight** of an object is the force with which gravity pulls it toward the earth:

$$w = mg$$

where w = weight, m = mass, and g = acceleration of gravity (9.8 m/s^2).

B. In the SI, mass rather than weight is normally specified.

C. On earth, the weight of an object (but not its mass) can vary because the pull of gravity is not the same everywhere on earth.

2-11. Third Law of Motion

A. The **third law of motion** states: When one object exerts a force on a second object, the second object exerts an equal force in the opposite direction on the first object.

B. No force ever occurs singly; for every **action force** there is an equal but opposite **reaction force**:

   1. The action force is the force the first object exerts on the second.
   2. The reaction force is the force the second object exerts on the first.

2-12. Circular Motion

A. **Centripetal force** is the inward force exerted on an object to keep it moving in a curved path:

$$F_c = \frac{mv^2}{r}$$

where F_c = centripetal force, m = mass, v = speed, and r = radius of circular path.

   1. The greater the object's mass, the greater the centripetal force.
   2. The faster the object moves, the greater the centripetal force.
   3. The smaller the circle (curved path) of the object, the greater the centripetal force.

B. Highway curves are **banked** so that the horizontal component of the reaction force the tilted road exerts on the car provides the centripetal force needed to keep the car from skidding.

2-13. Newton's Law of Gravity

A. Newton used Galileo's work on falling bodies and Kepler's three laws of gravity to develop his **law of gravity**.

B. Newton's law of gravity states: Every object in the universe attracts every other object with a force proportional to both of their masses and inversely proportional to the square of the distance between them.

$$F = \frac{Gm_1m_2}{R^2}$$

where F = force, G = gravitational constant (6.670×10^{-11} N • m^2/kg^2), and m_1 = mass of object 1, m_2 = mass of object 2, and R = distance between objects.

C. The force of gravitation drops off rapidly as the distance between two objects increases.

D. The **center of mass** of an object is the point where the mass of the object appears to be concentrated; for a uniform sphere the center of mass is its geometric center.

2-14. Artificial Satellites

A. The world's first artificial satellite was Sputnik I, launched in 1957 by the Soviet Union.
B. A satellite in a geostationary orbit remains in place indefinitely over a particular location on the earth.
C. To maintain a circular orbit, the centripetal force on the satellite must be equal to the gravitational force on the satellite.
D. The **escape speed** is the speed required by an object to leave the gravitational influence of an astronomical body; for earth this speed is about 40,000 km/h.

KEY TERMS AND CONCEPTS

The questions in this section will help you review the key terms and concepts from Chapter 2.

Multiple Choice

Circle the best answer for each of the following questions.

1. The rate at which an object travels a certain distance is known as
   a. velocity
   b. speed
   c. momentum
   d. acceleration

2. While driving between Pittsburgh and Chicago, the driver of the car looks at the car's speedometer and notes a speed of 50 miles/hour. The speedometer is indicating the car's
   a. average speed
   b. velocity
   c. terminal speed
   d. instantaneous speed

3. Which one of the following represents a vector quantity?
   a. 750 miles/h
   b. 60 cycles/second
   c. 220 volts
   d. 55 km/h toward the north

4. The acceleration of gravity at the earth's surface is
   a. 32 m/s^2
   b. 9.8 m/s^2
   c. 1.6 m/s^2
   d. 5.4 m/s^2

5. Newton's first law of motion deals with
   a. force
   b. acceleration
   c. weight
   d. inertia

6. A tennis player hits a tennis ball with a tennis racquet. If the force supplied by the racquet to the ball is the action force, what would be the reaction force?
   a. the force transmitted from the racquet to the tennis player's arm
   b. the resistance of air molecules on the tennis ball
   c. the force exerted by the tennis ball on the racquet
   d. the force exerted by the tennis ball on the opposing player's racquet

7. A bowling ball and a marble are dropped at the same time in a vacuum chamber. Which one of the following statements correctly describes the results?
   a. The bowling ball will land first.
   b. The acceleration of the bowling ball is greater than that of the marble.
   c. Both have the same speed upon landing.
   d. The force of impact will be the same for both the bowling ball and the marble.

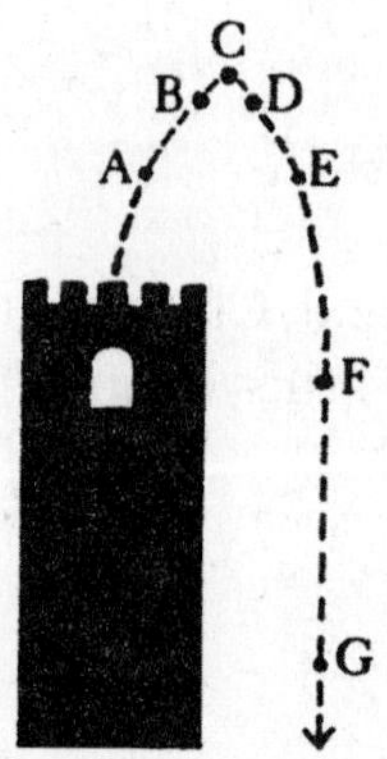

Refer to the above drawing to answer questions 8 and 9. A ball is thrown upward and is shown at several positions along its trajectory.

8. The speed of the ball at point C is
   a. greater than at point A
   b. at its maximum
   c. zero
   d. 9.8 m/s

9. The time interval between each change in position of the ball is
   a. greatest between points F and G
   b. at a minimum between points B and C and points C and D
   c. shortest between points F and G
   d. approximately equal

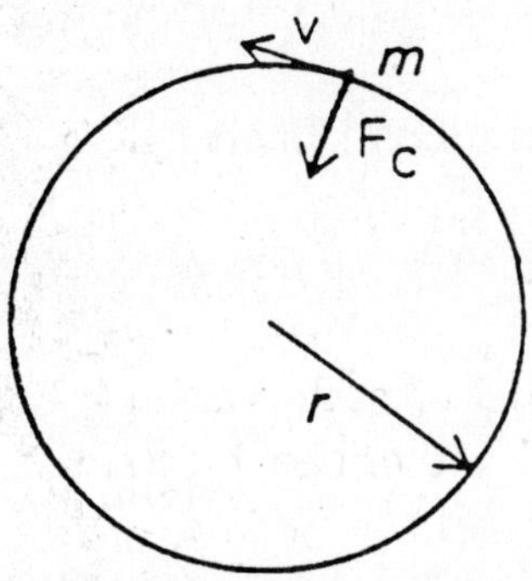

Refer to the above drawing to answer questions 10 through 12.

10. Given that F_c = centripetal force, v = speed, m = mass, and r = radius, doubling the value of m would
    a. reduce the value of F_c by 1/2
    b. double the value of F_c
    c. quadruple the value for F_c
    d. have no effect on the value of F_c

11. Doubling the value of v would
    a. reduce the value of F_c by 1/2
    b. quadruple the value of F_c
    c. double the value of F_c
    d. have no effect on the value of F_c

12. Doubling the value of r would
    a. double the value of F_c
    b. reduce the value of F_c by 1/2
    c. quadruple the value of F_c
    d. have no effect on the value of F_c

Refer to the above drawing to answer questions 13 and 14.

13. The string
    a. provides gravitational force
    b. represents the force of inertia
    c. represents the center of mass
    d. provides centripetal force

14. The centripetal force vector $\mathbf{F}_c$ is always
    a. tangent to the circle
    b. directed toward the center of the circle
    c. directed vertically toward the earth
    d. directed away from the center of the circle

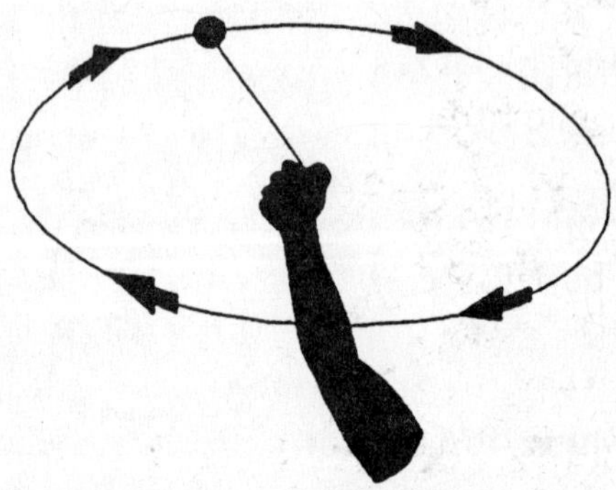

Refer to the above drawing to answer question 15. The car was originally at rest.

15. The best explanation for the position of the passengers within the car is
    a. the car's sudden start throws the passengers back into their seats
    b. the passenger's inertia tends to keep them in their original positions as the car (and the seats) move forward
    c. the car's forward acceleration is balanced by the opposing force of the passengers moving backward in their seats
    d. the car's seats broke at the moment of acceleration

True or False

Decide whether each statement is true or false. If false, briefly state why it is false or correct the statement to make it true. The first question has been completed as an example.

FALSE 1. Speed is an example of a vector quantity.

The speed at which an object moves reveals how fast the object is going, not its direction.

T 2. The SI unit of force is the newton.

T 3. Scalar quantities require only a number and a unit to be completely specified.

T 4. Assuming there is no wind or air friction, a ball thrown upward at a certain speed will return to its starting point with the same speed.

T 5. The speed or direction of any moving object stays unchanged unless a net force acts upon the object.

F 6. Inertia is the force that resists any change in an object's state of rest or motion.

T 7. The direction of acceleration of a moving object is always the same as the direction of the net force(s) acting on the object.

F 8. Your weight remains constant no matter where you are on the earth's surface.

F 9. The force that has to be applied to an object to make it move in a curved path is called centrifugal force.

T 10. If the distance between a planet and its sun were tripled, the gravitational force applied on the planet by its sun would drop to one-ninth of its original amount.

Fill in the Blank

Complete the following statements by filling in the missing term.

1. Speed is equal to distance divided by ___time___.

2. Distance is equal to the product of ___speed___ and time.

3. ___Time___ is equal to distance divided by speed.

4. According to Newton's second law of motion, acceleration is equal to ___Force___ divided by mass.

5. Force is equal to mass times ___acceleration___.

6. ___Force___ is equal to mass times the acceleration of gravity.

7. The curved path that an object takes when it is thrown upward at an angle to the ground is called a ___parabola___.

8. A ___vector___ may be represented by an arrowed line that represents the magnitude and direction of a quantity.

9. The ___pound___ is the British system unit of force.

10. The vector quantity that includes both speed and direction is called ___velocity___.

Matching

Match the terms on the left with their definitions on the right.

1.______ force
2.______ mass
3.______ Newton's law of gravity
4.______ inertia
5.______ vector quantity
6.______ Newton's first law of motion
7.______ newton
8.______ centripetal force
9.______ scalar quantity
10.______ Newton's third law of motion

a. inward force on an object moving in a curved path
b. resistance an object offers to any change in its state of rest or motion
c. states that, if no net force act on it, every object continues in its state of rest or uniform motion in a straight line.
d. states that when one object exerts a force on a second object, the second object exerts an equal but opposite force on the first.
e. has both magnitude and direction
f. states that every object in the universe attracts every other object with a force directly proportional to both their masses and inversely proportional to the square of the distance between them.
g. any influence that can cause an object to be accelerated.
h. has magnitude only
i. SI unit for power
j. quantity of matter in an object

SOLVED PROBLEMS

Study the following solved example problems as they will provide insight into solving the problems listed at the end of Chapter 2 in *The Physical Universe*. Review the mathematics refresher in the text if you are unfamiliar with the basic mathematical operations presented in these examples.

Example 2-1

You are standing 100 m north of your car when an alligator appears 20 m north of you and moves toward you at 8 m/s. At the same moment, you start to run toward your car at 5 m/s. Will you reach the car before the alligator reaches you?

Solution

Time is of the essence here! Both you and the pursuing alligator are moving toward the car. Because the alligator was 20 m north of you, it has to cover a distance of 120 m to reach the car.

To determine how long it will take the alligator to reach the car, we use the formula for time.

$$t = \frac{d}{v}$$

Substituting for *d* and *v* and dividing, we get

$$t = \frac{120 \text{ m}}{8 \text{ m/s}} = 15 \text{ s}$$

Note that the m units cancel, leaving the answer expressed in units of time (s).

Now let's find out how long it takes you to reach the car and safety.

$$t = \frac{d}{v} = \frac{100 \text{ m}}{5 \text{ m/s}} = 20 \text{ s}$$

You lose! The alligator overtakes you on the way to the car.

Example 2-2

A 1240-kg car goes from 16 m/s to 26 m/s in 20 s. What is the average force acting upon it?

Solution

To determine the average force acting on the car, we use Newton's second law of motion expressed as:

$$F = ma$$

We are given the car's mass (1240 kg), but not its acceleration. To determine the car's acceleration, we can use the formula for acceleration:

$$a = \frac{v_f - v_i}{t}$$

Substituting for v_f, v_i, and *t*, we get:

$$a = \frac{26 \text{ m/s} - 16 \text{ m/s}}{20 \text{ s}} = \frac{10 \text{ m/s}}{20 \text{ s}} = 0.5 \text{ m/s}^2$$

We can now substitute the quantities for mass and acceleration into the formula for force and solve the problem.

$$F = ma = (1240\text{ kg})(0.5\text{ m/s}^2) = 620\text{ kg} \bullet \text{m/s}^2 = 620\text{ N}$$

Example 2-3

A traditional Scottish highland game is the hammer throw in which the athlete spins the hammer in a circular motion building momentum prior to releasing the hammer. If the head of the hammer weighs 6.7 kg and is moving at 6 m/s in a circular arc of 1.2 m radius, how much force must the athlete exert on the handle of the hammer to prevent it from flying out of his hands? Ignore the mass of the hammer's light wood handle.

This is a centripetal force problem and to solve it, we use the formula for centripetal force.

$$F_c = \frac{mv^2}{r}$$

Substituting the quantities given to us in the problem, we get

$$F_c = \frac{mv^2}{r} = \frac{(6.7\text{ kg})(6\text{ m/s})^2}{1.2\text{ m}} = \frac{241.2\text{ kg} \bullet \text{m}^2/\text{s}^2}{1.2\text{ m}} = 201\text{ N}$$

WEB LINKS

This interactive web site illustrates the concept of velocity composition by showing the velocity vectors relevant to the motion of a boat on a river. Requires the Macromedia Shockwave Player.

http://users.erols.com/renau/velocity_composition.html

This interactive web site displays the acceleration and velocity vectors in real time of a puck in motion. Requires the Macromedia Shockwave Player.

http://users.erols.com/renau/velocity_applet.html

Control the trajectories of a cannonball at this interactive web site

http://galileo.phys.virginia.edu/classes/109N/more_stuff/Applets/newt/applet.html

Study projectile orbits and satellite orbits at this interactive web site

http://www.phy.ntnu.edu.tw/java/projectileOrbit/projectileOrbit.html

ANSWER KEY

Multiple Choice

1. b 2. d 3. d 4. b 5. d 6. c 7. c 8. c 9. d 10. b 11. b 12. b 13. d 14. b 15. b

True or False

1. False. See explanatory notes given in the example.
2. True
3. True
4. True
5. True
6. False. Inertia is not an actual force. A force is any influence that can change the speed or direction of motion of an object. Inertia is the resistance offered by an object to any change in its state of rest or motion.
7. True
8. False. The pull of gravity is not exactly the same everywhere on earth; therefore, a person's weight will vary depending upon the person's location.
9. False. The force that has to be applied to make an object move in a curved path is called centripetal force.
10. True

Fill in the Blank

1. time
2. speed
3. time
4. force
5. acceleration
6. weight
7. parabola
8. vector
9. pound
10. velocity

Matching

1. g 2. j 3. f 4. b 5. e 6. c 7. i 8. a 9. h 10. d

Chapter 3

ENERGY

OUTLINE

Work
3.1 The Meaning of Work
3.2 Power

Energy
3.3 Kinetic Energy
3.4 Potential Energy
3.5 Energy Transformations
3.6 Conservation of Energy
3.7 The Nature of Heat

Momentum
3.8 Linear Momentum
3.9 Rockets
3.10 Angular Momentum

Relativity
3.11 Special Relativity
3.12 Rest Energy
3.13 General Relativity

Energy and Civilization
3.14 The Energy Problem
3.15 The Future

Goals

1. Explain what is meant by work.
2. Explain what is meant by power.
3. Distinguish between kinetic energy and potential energy.
4. Give several examples of potential energy.
5. State the law of conservation of energy and give several examples of energy transformations.
6. Discuss why heat is today regarded as a form of energy rather than as an actual substance.
7. Define linear momentum and discuss its significance.
8. Use the principle of conservation of linear momentum to analyze the motion of objects that push each other apart, for instance, when a rocket is fired.
9. State what is meant by angular momentum.
10. Explain how conservation of angular momentum is used by skaters to spin faster and by footballs to travel farther.
11. Describe several relativistic effects and indicate why they are not conspicuous in the everyday world.
12. Explain what is meant by rest energy and be able to calculate the rest energy of an object of given mass.
13. Describe how gravity is interpreted in Einstein's general theory of relativity.
14. Discuss the various factors that are part of the energy problem of the future.

CHAPTER SUMMARY

Chapter 3 begins by defining **work** and **power**. The difference between **kinetic** and **potential energy** is explained, and the concept of energy transformation is introduced. **Heat** is shown to be a form of energy. The **law of conservation of energy** is presented, and mass and energy are stated to be equivalent. The discussion of Einstein's **special theory of relativity** shows that measurements of length, time, and mass depend upon the relative velocity between the observer and what is being observed. The discussion of Einstein's **general theory of relativity** explains that gravity results from the distortion of space-time around a body of matter. The environmental and technical problems of providing enough energy to meet the demands of the growing global population are presented, and alternative energy sources are discussed.

CHAPTER OUTLINE

3-1. The Meaning of Work

A. **Work** equals force times distance.

$$W = Fd$$

1. When a force is applied to an object, the object must move a distance or no work is done.
2. When the force is not parallel to the distance d, the projection of the force in the direction of d must be used to find the work.
3. No work is done when a force is perpendicular to the direction of motion of an object.

B. The SI unit of work is the **joule**.

$$1 \text{ joule (J)} = 1 \text{ newton-meter (N} \bullet \text{m)}$$

C. Work done against gravity is represented by:

$$W = mgh$$

where W = work, mg = mass times the acceleration of gravity (9.8 m/s^2), and h = total height.

3-2. Power

A. **Power** is the rate at which work is being done:

$$P = \frac{W}{t}$$

where P = power, W = work, and t = time.

B. The SI unit of power is the **watt**.

$$1 \text{ watt (W)} = 1 \text{ joule/second (J/s)}$$

C. The **kilowatt** (kW) is a convenient unit of power for many applications.

3-3. Kinetic Energy

A. **Energy** is that property something has that enables it to do work.

B. The energy of a moving object is called **kinetic energy** (KE):
where m = mass and v = speed.

$$\text{KE} = \tfrac{1}{2}mv^2$$

1. Since KE is the product of mass and the square of the speed, the greater the mass and/or speed the greater the KE.
2. KE increases very rapidly with speed because of the v^2 factor.

3-4. Potential Energy

A. **Potential energy** (PE) is the energy an object has by virtue of its position.

B. The amount of work an object performs as it moves toward the earth under the influence of gravity is its gravitational potential energy:

$$\text{PE} = mgh$$

where m = mass, g = acceleration of gravity (9.8 m/s^2), and h = height.

C. The gravitational PE of an object is a relative quantity because it depends on the level from which it is reckoned.

3-5. Energy Transformations

A. All forms of energy can be transformed or converted from one form to another; most events in the physical world involve energy transformation.

B. Energy exists in a variety of forms. Some forms other than kinetic and potential include:
   1. Chemical energy
   2. Heat energy
   3. Electric energy
   4. Radiant energy

3-6. Conservation of Energy

A. The **law of conservation of energy** states that energy cannot be created or destroyed, although it can be changed from one form to another.
B. Matter can be considered as a form of energy; matter can be transformed into energy and energy into matter according to the law of conservation of energy.

3-7. The Nature of Heat

A. Heat was once thought to be a weightless, invisible, odorless, and tasteless substance called **caloric**.
B. James Prescott Joule was the first to demonstrate conclusively that heat is energy.

3-8. Linear Momentum

A. **Linear Momentum** is a measure of the tendency of a moving object to continue in motion along a straight line:

$$\mathbf{p} = m\mathbf{v}$$

where $\mathbf{p}$ = linear momentum, m = mass, and $\mathbf{v}$ = velocity.
B. The greater an object's linear momentum, the harder it is to change the object's velocity, that is, its speed and/or direction of motion.
C. The **law of conservation of momentum** states: In the absence of outside forces, the total momentum of a set of objects remains the same no matter how the objects interact with one another.

3-9. Rockets

A. The operation of a rocket is based on conservation of momentum.
   1. The momentum of the exhaust gases is balanced by the rocket's upward momentum.
   2. The total momentum of the entire system is zero.
B. **Multistage rockets** are more efficient than single-stage, and so are widely used.

3-10. Angular Momentum

A. **Angular momentum** is a measure of the tendency of a rotating object to continue spinning about a fixed axis; **conservation of angular momentum** is the description of the tendency of spinning objects to remain spinning.

1. The greater the mass of an object and the more rapidly it rotates, the greater its angular momentum.
2. The angular momentum of a spinning object also depends upon how its mass is distributed; the farther away from the axis of rotation its mass is distributed, the more the angular momentum.

3-11. Special Relativity

A. Albert Einstein (1879-1955) published the **special theory of relativity** in 1905.

B. Special relativity is based on two postulates:

1. The laws of physics are the same in all **frames of reference** moving at constant velocity.
2. The speed of light (c) in free space has the same value for all observers ($c = 3 \times 10^8$ m/s)

C. According to the special theory of relativity, when there is relative motion between an observer and what is being observed, lengths are observed to be shorter than when at rest, time intervals are observed to be longer, and masses are observed to be greater.

D. Nothing material can travel as fast or faster than the speed of light.

3-12. Rest Energy

A. The **rest energy** of a body is the energy equivalent of its mass:

$$E_O = m_O c^2$$

where E_O = rest energy, m_O = rest mass, and c = speed of light.

B. Rest energy is liberated in all energy-producing reactions of chemistry and physics.

3-13. General Relativity

A. Einstein published the **general theory of relativity** in 1916, which related gravitation to the structure of space and time and showed that even light was subject to gravity.

B. Einstein spent the last years of his life unsuccessfully trying to join together the phenomena of gravitation and electromagnetism into a "unified field theory."

3-14. The Energy Problem

A. Except for nuclear energy and heat energy from within the earth, all available energy is directly or indirectly derived from the sun.
B. Coal, oil, and natural gas are called fossil fuels because they were formed from the remains of organisms that lived millions of years ago.
C. The use of fossil fuels presents several disadvantages.
    1. Oil and natural gas reserves will last about another century..
    2. Although coal reserves will last several hundred more years, mining coal is dangerous, and burning coal creates environmental problems such as acid rain, air pollution, and enhanced global warming.
D. Nuclear energy presents both advantages and problems.
    1. Nuclear fuel reserves exceed those of fossil fuels, especially with the development of "breeder" reactors.
    2. Properly built and run nuclear power plants are excellent energy sources.
    3. The potential for a large-scale nuclear accident is present.
    4. Discharge of radioactive wastes into the environment from badly run nuclear power plants has occurred.
    5. An unsolved disposal problem of radioactive nuclear waste exists.

3-15. The Future

A. Nuclear fusion represents a potentially safe, nonpolluting, and inexhaustible energy source; however, the problem is how to manage until nuclear fusion arrives.
B. Alternate energy sources, such as solar energy, present various technical and environmental problems; some are not economically feasible.
C. Wise use of alternate energy sources, conservation of energy, and a slowdown in human population growth are necessary to avoid disaster until nuclear fusion becomes a reality.

KEY TERMS AND CONCEPTS

The questions in this section will help you review the key terms and concepts from Chapter 3.

Multiple Choice

Circle the best answer for each of the following questions.

1. The SI unit of work is the
    a. joule
    b. watt
    c. calorie
    d. newton

2. The SI unit of energy is named after
   a. Albert Einstein
   b. James Prescott Joule
   c. Count Rumford
   d. James Watt

3. The property that something has that enables it to do work is
   a. mass
   b. energy
   c. momentum
   d. force

4. A billiards player strikes the cue ball with a cue stick, thereby putting the cue ball in motion. The cue ball will continue to move until
   a. it runs out of speed
   b. its inertia is balanced by its momentum
   c. its linear momentum is converted into angular momentum
   d. the forces of friction and air resistance bring it to a stop

5. The same billiards player "breaks" by striking the cue ball with the cue stick and sending it into a stationary formation of billiard balls, thereby putting the balls in motion. The law of conservation of momentum tells us
   a. the total momentum of the cue ball and the billiard balls has been reduced
   b. it is impossible for the momentum of the billiard balls to be changed
   c. the total combined momentum of the cue ball and the billiard balls remains unchanged after the break
   d. the momentum of the cue ball is reduced by a factor equal to the number of billiard balls put into motion

6. A spinning top eventually falls over and comes to rest because
   a. it has lost all its angular momentum
   b. it has lost all of its linear momentum
   c. the force of gravity has brought it to a stop
   d. it has lost all its original inertia

7. In the cylinder of an automobile engine, the chemical energy of gasoline is initially transformed into
   a. mechanical energy
   b. heat energy
   c. electrical energy
   d. potential energy

8. The wound spring of a watch possesses _______________ energy.

a. potential
b. kinetic
c. chemical
d. zero

9. A 75-kg window air conditioner falls a distance of 6 m to the street below (no one is injured). What was the PE of the air conditioner while it sat at rest within the window frame?

a. 735 J
b. 2205 J
c. 3601.5 J
d. 4410 J

10. A 1000-kg car is traveling at a speed of 10 m/s. If the speed of the car is increased to 20 m/s, the car's kinetic energy is

a. doubled
b. halved
c. quadrupled
d. unchanged

11. The law of conservation of energy states:

a. mass is a form of energy
b. it is illegal to waste energy
c. the end product of all energy transformation is heat
d. energy transformations occur without a net gain or loss of energy

12. Einstein's general theory of relativity

a. relates the movement of small particles in a fluid to their bombardment by molecules
b. describes light as having both wave and particle properties
c. interprets gravity as a distortion in the structure of space and time
d. proposed that heat was a form of energy

13. A spaceship and its crew travel at near the speed of light to reach the star Proxima Centauri. Observers on earth note that it takes 4.3 years for the crew to reach Proxima Centauri. To the spaceship crew the trip takes

a. more than 4.3 years
b. less than 4.3 years
c. exactly 4.3 years
d. zero time

14. The distance from Earth to Proxima Centauri is approximately 40.7 trillion kilometers as measured by earthbound observers. From the frame of reference of the spaceship crew, the distance to Proxima Centauri is _______________ and the clock on board the spaceship is running _______________ than clocks on earth.
    a. shorter; faster
    b. shorter; slower
    c. longer; faster
    d. longer; slower

15. All are advantages of solar cells except
    a. little or no maintenance expense
    b. no moving parts
    c. can be installed close to where electricity is needed
    d. generate electricity more cheaply than fossil fuel plants for a given power output

True or False

Decide whether each statement is true or false. If false, either briefly state why it is false or correct the statement to make it true. See previous chapters for an example.

__T__ 1. Farmer Ben attempts to remove an 80-kg rock from his field. If he applies a force of 789 N for 10 seconds yet fails to move the rock, he has done no work.

__F__ 2. Upon eating food, the chemical energy contained within the food is destroyed by the process of digestion.

__T__ 3. In a nuclear weapon, a small amount of matter is converted into energy.

__F__ 4. Energy can be defined as the rate at which work is being done.

__T__ 5. Kinetic energy may be thought of as the energy of motion.

__F__ 6. Your car runs out of gas along a level highway. If the car's mass is 1000 kg and you push the car 20 m to the nearest gas station, the work you have done is the product of the car's mass (1000 kg) and distance the car was pushed (20 m).

F ____ 7. According to Einstein's special theory of relativity, the speed of light in free space has the same value for all observers.

F ____ 8. Einstein's general theory of relativity predicts that light is subject to gravity.

F ____ 9. Coal reserves are expected to be depleted within the next few decades.

F ____ 10. Nuclear energy accounts for about 25 percent of the world's energy consumption.

Fill in the Blank

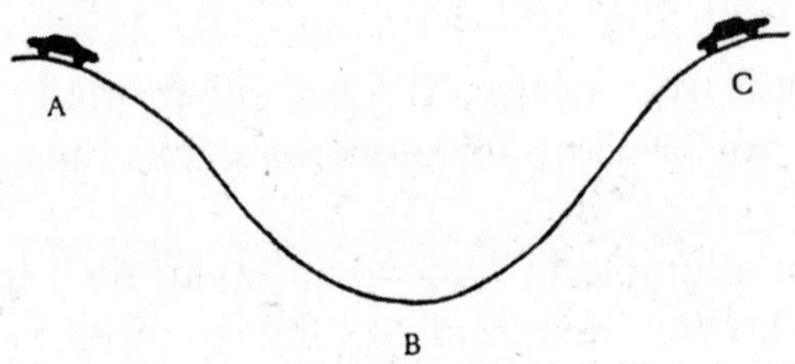

In the above drawing, the car coasts from point A to point C (assume friction is not acting on the car). Refer to this drawing to answer questions 1 through 4.

1. As the car coasts from point A to point B, its initial potential energy is converted into __kinetic__ energy.

2. The kinetic energy of the car is greatest at point __B__.

3. The kinetic energy of the car is totally converted into potential energy at point __C__.

4. As the car coasts from point A to point C, the total amount of energy (KE + PE) is __unchanged__.

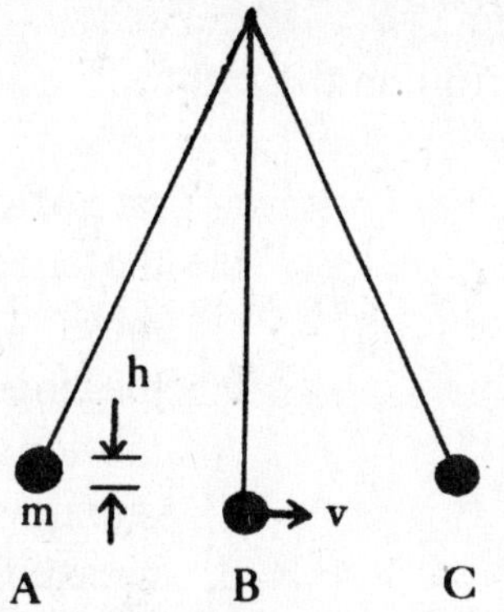

In the above drawing, the pendulum ball is moving to the right. Refer to this drawing to answer questions 5 and 6.

5. The pendulum ball has minimum potential energy and maximum kinetic energy at point ____B____.

6. As the pendulum ball moves from point A to point B, the ____pot____ energy of the ball is transformed to ____kinetic____ energy.

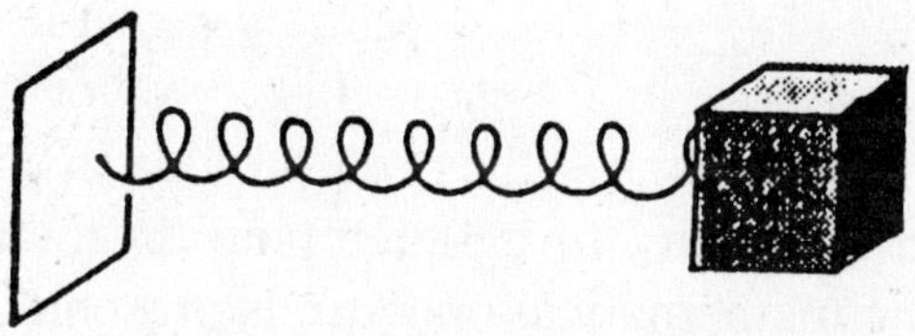

Stretched spring

7. In the above drawing, the stretched spring has ____pot.____ energy.

8. The potential energy of a raised object is equal to the ____work____ used to lift it.

9. According to general relativity, gravity is a warping of the structure of space and time due to the presence of ____matter____.

10. The kinetic energy of a moving object is proportional to the square of its ____speed____.

Matching

Match the term on the left with its definition on the right.

1._____ fossil fuels

2._____ sun

3._____ power

4._____ work

5._____ rest energy

6._____ joule

7._____ angular momentum

8._____ potential energy

9._____ kinetic energy

10._____ linear momentum

a. rate of doing work
b. the energy an object has by virtue of its mass
c. the energy an object has by virtue of its motion
d. one newton-meter
e. measure of the tendency of a rotating object to continue spinning about a fixed axis
f. force times distance
g. the energy an object has by virtue of its position
h. coal, oil, and natural gas
i. measure of the tendency of a moving object to continue motion along a straight line
j. original source of most energy on earth

SOLVED PROBLEMS

Study the following solved example problems as they will provide insight into solving the problems listed at the end of Chapter 3 in *The Physical Universe*. Review the mathematics refresher in the text if you are unfamiliar with the basic mathematical operations presented in these examples.

Example 3-1

A total of 784 J of work is needed to lift a metal chair a height of 10 m. What is the mass of the chair?

Solution

The work W needed to raise an object of mass m to a height h above its original position is

$$W = mgh$$

To determine the mass of the body, we must solve for m by dividing both sides of the equation by gh:

$$m = \frac{W}{gh}$$

Substituting the quantities given in the problem and the acceleration of gravity for g, we have:

$$m = \frac{784 \text{ J}}{(9.8 \text{ m/s}^2)(10 \text{ m})}$$

The answer must be expressed in kg. To do this we must use the definitions of joule, newton-meter, and newton.

$$m = \frac{784 \text{ N} \bullet \text{m}}{(9.8 \text{ m/s}^2)(10 \text{ m})}$$

$$m = \frac{784 \text{ kg} \bullet \text{m/s}^2 \bullet \text{m}}{(9.8 \text{ m/s}^2)(10 \text{ m})}$$

$$m = \frac{784 \text{ kg}}{98}$$

$$m = 8 \text{ kg}$$

The mass of the chair is 8 kg.

Example 3-2

A 80-kg person hikes to the top of Diamond Head near Honolulu, Hawaii. If the vertical gain in elevation is 171 m and it took 60 minutes to reach the top, how much power did the person develop?

Solution

The formula for power is:

$$P = \frac{W}{t}$$

We must first find the value for W, the work done by the person in climbing 171 m. In Chapter 3, we learned that the work needed to raise an object of mass m to a height h above its original position is

$$W = mgh$$

where g is the acceleration of gravity (9.8 m/s^2). Substituting the quantities for m and h into the formula for W we get

$$W = (80 \text{ kg})(9.8 \text{ m/s}^2)(171 \text{ m})$$

$$W = 134{,}064 \text{ kg} \bullet \text{m/s}^2 \bullet \text{m}$$

$$W = 134{,}064 \text{ N} \bullet \text{m}$$

$$W = 134{,}064 \text{ J}$$

We can now determine P by substituting the quantities for W and t into the formula for power.

Note that the SI unit of power is the watt (W), where

1 watt (W) = 1 joule/second (J/s)

We must convert time in minutes as given in the problem into time in seconds.

$$(60 \text{ min.})\left(\frac{60 \text{ s}}{1 \text{ min.}}\right) = 3600 \text{ s}$$

The power the person developed is:

$$P = \frac{W}{t} = \frac{134{,}064 \text{ J}}{3600 \text{ s}} = 37.2 \text{ J/s} = 37 \text{ W}$$

WEB LINKS

Participate in an interactive demonstration of linear momentum at

http://library.thinkquest.org/3042/java/linear_demo.html

Play a game that illustrates potential and kinetic energy at

http://library.thinkquest.org/3042/java/energy_demo.html

Take a virtual spaceflight and observe the effect of Special Relativity at

http://webphysics.ph.msstate.edu/jc/library/25-5/simulation.html

Learn about renewable and alternate energy sources at

http://solstice.crest.org/

ANSWER KEY

Multiple Choice

1. a 2. b 3. b 4. d 5. c 6. a 7. b 8. a 9. d 10. c 11. d 12. c 13. b 14. a 15. d

True or False

1. True
2. False. According to the law of conservation of energy, energy cannot be created or destroyed.
3. True
4. False. Power is the rate at which work is being done.
5. True
6. False. Work would be calculated by the formula $W = Fd$. Since the amount of force needed to move the car is not given, the amount of work cannot be determined. The car's weight ($w = mg$) does not matter since the downward pull of gravity is a force acting perpendicular to the direction of motion and can do no work on the car.
7. True
8. True
9. False. Coal reserves are expected to last for several centuries.
10. False. Nuclear energy accounts for about 6 percent of the world's energy consumption.

Fill in the Blanks

1. kinetic
2. B
3. C
4. unchanged
5. B
6. potential, kinetic
7. potential
8. work
9. matter
10. speed

Matching

1. h 2. j 3. a 4. f 5. b 6. d 7. e 8. g 9. c 10. i

Chapter 4

MATTER AND HEAT

OUTLINE

GOALS

1. Distinguish between temperature and heat.
2. Describe how various thermometers work.
3. Convert temperatures from the celsius to the fahrenheit scale and vice versa.
4. Discuss what is meant by the metabolic rate of an animal and how the maximum rate is related to the animal's size.
5. Define density and calculate the mass of a body of matter given its density and volume.
6. Define pressure and account for the increase in pressure with depth in a liquid or gas.
7. State Archimedes' principle and explain its origin.
8. Use Boyle's law to relate pressure and volume changes in a gas at constant temperature.
9. Use Charles's law to relate temperature and volume changes in a gas at constant pressure.
10. Explain the significance of the absolute temperature scale and the meaning of absolute zero.
11. Show how the ideal gas law is related to Boyle's law and Charles's law.
12. State the three basic assumptions of the kinetic theory of gases.
13. Discuss the connection between temperature and molecular motion.
14. Account for the different of gases, liquids, and solids in terms of the forces between their molecules.
15. Distinguish between evaporation and boiling.
16. Explain what is meant by heat of vaporization and by heat of fusion.
17. Discuss why heat engines cannot be perfectly efficient.
18. Compare heat engines and refrigerators.
19. State the two laws of thermodynamics.
20. Relate entropy to the second law of thermodynamics.

CHAPTER SUMMARY

In Chapter 4, **heat** is interpreted as molecular kinetic energy, and the connection between heat and the kinetic energies of atoms and molecules is discussed. **Temperature** is interpreted as a measure of average molecular kinetic energy, and the **celsius scale** and the **absolute temperature scale** are introduced. The relationship between a living organism's **metabolism** and work output is examined. The behavior of gases is summarized by the gas laws and the kinetic theory of gases. Changes of state between gases and liquids, and between liquids and solids, are discussed within the overall context of a **kinetic theory of matter**. Heat transformation into mechanical energy or work and the operation and efficiency of **heat engines** are explained by the laws of **thermodynamics**. The concept of **entropy** is discussed and related to the ultimate fate of the universe.

CHAPTER OUTLINE

4-1. Temperature

A. Temperature is that property of matter that gives rise to the sensations of hot and cold.
B. A **thermometer** is a device that measures temperature.
C. A **thermostat** makes use of the different rates of thermal expansion in the metals of a bimetallic strip to switch heating and cooling systems on and off.
D. Two temperature scales are used in the United States:
   1. The **fahrenheit scale** on which water freezes at 32°F and boils at 212°F at sea level.
   2. The **celsius scale** on which water freezes at 0°C and boils at 100°C at sea level.

4-2. Heat

A. The **heat** in a body of matter is the sum of the kinetic energies of all the separate particles that make up the body; this heat content is also called **internal energy**.
B. The SI unit of heat is the joule.
C. The **specific heat capacity** (or **specific heat**) of a substance is the amount of heat that must be added or removed from 1 kg of the substance in order to change its temperature by 1°C.
D. Heat can be transferred in three ways:
   1. **Conduction**, in which heat is transferred from one place to another by molecular collisions.
   2. **Convection**, in which heat is carried by the motion of a volume of hot fluid.
   3. **Radiation**, in which heat is transferred by electromagnetic waves.

4-3. Metabolic Energy

A. The complex of biochemical reactions that make food energy available for use by living organisms is called **metabolism**.

B. A **kilocalorie** is the amount of heat needed to change the temperature of 1 kg of water by 1°C; it is equal to one dietary "calorie."
C. The conversion of metabolic energy into biological work is relatively inefficient; much of the energy is lost as heat.

4-4. Density

A. The **density** of a substance is its mass per unit volume:

$$d = \frac{m}{V}$$

where d = density, m = mass of substance, and V = unit volume.
B. The proper SI unit of density is the kg/m^3; however, densities are often given in g/cm^3.

4-5. Pressure

A. The **pressure** on a surface is the perpendicular force per unit area:

$$p = \frac{F}{A}$$

where p = pressure, F = force, and A = area
B. The SI unit of force is the **pascal**:

$$1 \text{ pascal} = 1 \text{ Pa} = 1 \text{ newton/meter}^2$$

Because the pascal is a small unit, the **kilopascal** is often used (1 kPa = 1000 Pa).
C. **Hydraulic** machines are those in which forces are transmitted by liquids; **pneumatic** machines use compressed air.
D. Pressure in a fluid increases with depth.
   1. Atmospheric pressure at sea level averages 101 kPa (equals approximately 15 lb/in^2).
   2. Instruments called **barometers** measure atmospheric pressures.

4-6. Buoyancy

A. The upward force exerted on an object immersed in a fluid is called **buoyant force**

$$F_b = dVg$$

where F_b = buoyant force, d = density of the fluid, V = volume of the fluid displaced by the solid object, and g = the acceleration of gravity (9.8 m/s^2).

1. Buoyant force is always an upward force because the pressure underneath an immersed object is greater than the pressure above it.
2. **Archimedes' principle** states: Buoyant force on an object in a fluid is equal to the weight of fluid displaced by the object.

4-7. Gas Laws

A. **Boyle's law** states that the volume of a gas is inversely proportional to its pressure when the temperature is held constant:

$$\frac{p_1}{p_2} = \frac{V_2}{V_1} \quad \text{(at constant temperature)}$$

where p_1 = initial pressure, V_1 = initial volume, p_2 = final pressure, and V_2 = final volume.

B. **Absolute zero** is -273°C and is the theoretical but unreachable lowest possible temperature.

C. **Absolute temperatures** are temperatures in °C above absolute zero, denoted K in honor of the English physicist Lord Kelvin.

D. **Charles's law** states that the volume of a gas is directly proportional to its absolute temperature:

$$\frac{V_1}{V_2} = \frac{T_1}{T_2} \quad \text{(at constant pressure; temperatures on absolute scale)}$$

where V_1 = initial volume, T_1 = initial temperature (K), V_2 = final volume, and T_2 = final temperature (K).

E. The **ideal gas law** combines Boyle's and Charles's laws into a single formula:

$$\frac{p_1 V_1}{T_1} = \frac{p_2 V_2}{T_2} \quad \text{(temperatures on absolute scale)}$$

where p = pressure, V = volume, and T = temperature (K).

4-8. Kinetic Theory of Gases

A. Gas molecules are small compared with the average distance between them; a gas is mostly empty space.
   1. Gases are easily compressed.
   2. Gases are easily mixed.
   3. The mass of a certain volume of gas is much smaller than that of the same volume of a liquid or a solid.

B. Gas molecules collide without loss of kinetic energy.

C. Gas molecules exert almost no forces on one another, except when they collide.

4-9. Molecular Motion and Temperature

A. The absolute temperature of a gas is proportional to the average kinetic energy of its molecules.
B. Gas molecules, even at 0 K (-273°C), would still possess a small amount of kinetic energy.
C. Compression of a gas increases its temperature; expansion decreases its temperature.

4-10. Liquids and Solids

A. The intermolecular attractions between the molecules of a liquid are stronger than those in a gas but weaker than those in a solid.
B. Molecules of a solid do not move freely about but vibrate around fixed positions.

4-11. Evaporation and Boiling

A. Evaporation occurs when fast-moving molecules in a liquid leave the liquid's surface and escape into the air.
B. The **boiling point** of a liquid is that temperature at which the molecules of the liquid attain enough kinetic energy to overcome their intermolecular attractions, resulting in a change of state from liquid to gas.
C. Evaporation differs from boiling in two ways:
   1. Evaporation occurs only at a liquid surface; boiling occurs in the entire volume of liquid.
   2. Evaporation occurs at all temperatures; boiling occurs only at the boiling point.
D. The **heat of vaporization** of a substance is the amount of heat energy needed to change 1 kg of the substance from liquid to gas at its boiling point.

4-12. Melting

A. The **heat of fusion** of a solid is the amount of heat needed to melt 1 kg of it. It is also equal to the amount of heat given off by 1 kg of a liquid when it hardens into a solid.
B. **Sublimation** is the direct conversion of a substance from the solid to the vapor state, or from the vapor state to the solid state, without it entering the liquid state.

4-13. Heat Engines

A. A **heat engine** is a device that converts heat into mechanical energy or work.
B. During operation a heat engine extracts energy from a flow of heat through it.
   1. Some of the heat flow is converted into mechanical energy; the rest is lost as waste heat.
   2. To produce a flow of heat energy, the heat must travel from a hot reservoir to a cold reservoir.

C. In an automobile, the ignited gases within the cylinder constitute the hot reservoir and the atmosphere is the cold reservoir.
D. A **refrigerator** operates as the reverse of a heat engine by using mechanical energy to force heat to flow from a cold reservoir (the refrigerator interior) to a hot reservoir (the atmosphere).

4-14. Thermodynamics

A. **Thermodynamics** is the science of heat transformation.
B. The **first law of thermodynamics** states: Energy cannot be created or destroyed, but it can be converted from one form to another.
C. The **second law of thermodynamics** states: It is impossible to take heat from a source and change all of it to mechanical energy or work; some heat must be wasted.
D. Maximum efficiency of a heat engine depends on the temperatures at which it takes in and ejects heat; the greater the ratio between the two temperatures, the more efficient the engine:

$$\text{Eff}_{max} = 1 - \frac{T_{cold}}{T_{hot}} \quad (T = \text{temperature on absolute scale})$$

4-15. Fate of the Universe

A. The heat energy of the universe increases at the expense of other forms of energy.
B. "Heat death" of the universe will occur when all particles of matter ultimately have the same average kinetic energy and exist in a state of maximum disorder.

4-16. Entropy

A. **Entropy** is a measure of the disorder of the particles that make up a body of matter.
B. In terms of entropy the second law of thermodynamics becomes: The entropy of a system isolated from the rest of the universe cannot decrease.
C. Living organisms require a constant energy input to overcome entropy and maintain the order of their biological systems.
D. The entropy of the universe is increasing with the passage of time.

KEY TERMS AND CONCEPTS

The questions in this section will help you review the key terms and concepts from Chapter 4.

Multiple Choice

Circle the best answer for each of the following questions.

1. Bimetallic strips composed of two different metals each having a different rate of thermal expansion are used in
   a. pressure cookers
   b. thermostats
   c. heat engines
   d. pressure gauges

2. A temperature of 100°C is equal to what temperature on the fahrenheit scale?
   a. 0°
   b. -273°
   c. 212°
   d. 32°

3. A block of wood measuring 10 cm x 10 cm x 1 cm is dropped into a container of water and floats on the water's surface. A second block of wood having the same dimensions is dropped into the container and sinks. From these observations you know that
   a. the weight of the first block is less than the weight of the second block
   b. the weight of the first block is greater than the weight of the second block
   c. the weight of the first block is greater than the buoyant force acting upon it
   d. the weight of the second block is less than the buoyant force acting upon it

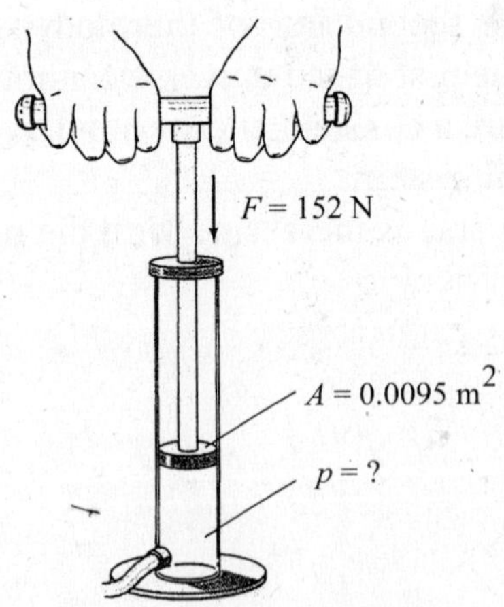

Refer to the above drawing to answer Question 4 on page 51.

4. A force of 152 N is applied to an area of 0.0095 m^2. What is the resulting pressure?
   a. 1.44 Pa
   b. 152.095 Pa
   c. 6.25×10^{-5} Pa
   d. 1.6×10^5 Pa

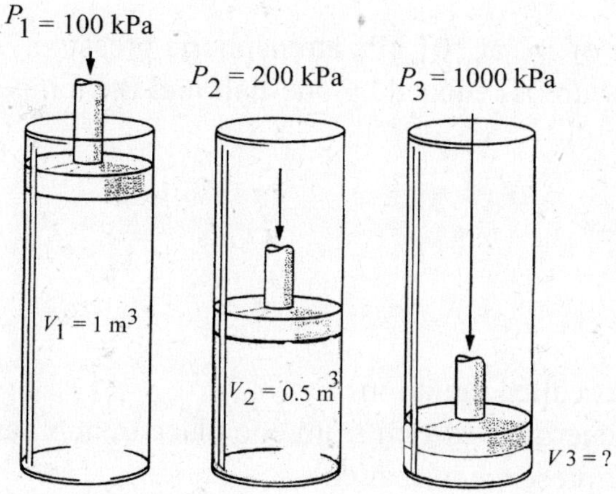

Refer to the above drawing to answer questions 5 and 6. In this drawing, a gas is being compressed within a cylinder. Here $P_1V_1 = P_2V_2 = P_3V_3 = 10^5$ N • m.

5. As the pressure is increased from 200 kPa to 1000 kPa, V_3 becomes
   a. 0.25 m^3
   b. 0.1 m^3
   c. 0.8 m^3
   d. 0.008 m^3

6. The drawing illustrates
   a. Charles's law
   b. Boyle's law
   c. the ideal gas law
   d. Archimedes' principle

7. According to Charles's law, as the absolute temperature of a gas increases from +273 K to +546 K at constant pressure, the volume of the gas
   a. remains constant
   b. decreases by one-half
   c. quadruples
   d. doubles

8. According to the ideal gas law, if both the volume and the absolute temperature of a gas are doubled, the pressure
   a. remains unchanged
   b. doubles
   c. is reduced by one-half
   d. is quadrupled

9. A balloon has a volume of 5 L at 101 kPa atmospheric pressure. If the balloon is placed in a chamber where the pressure is reduced by one-half and the temperature is held constant, the volume of the balloon will
   a. be reduced to 2.5 L
   b. be increased to 10 L
   c. be increased to 25 L
   d. remain unchanged

10. The type of heat transfer called radiation
   a. takes place as heat energy is carried from one place to another by molecular collisions
   b. does not require the presence of matter
   c. involves the motion of a volume of hot fluid
   d. occurs when a solid changes into its vapor state without first becoming a liquid

11. Absolute zero, or 0 K, is the lowest possible temperature because
   a. no scientific instrument has yet been designed that can measure temperatures lower than 0 K
   b. this is the lowest temperature ever recorded on earth or anywhere else in the universe
   c. entropy is at a maximum at this temperature
   d. there can be no smaller amount of molecular kinetic energy than that at 0 K

12. It is impossible to lower the temperature of a gas to absolute zero (assuming that such a low temperature can be reached) because
   a. all known gases turn into liquids before absolute zero is reached
   b. gases cannot be cooled below +273 K
   c. the gas would acquire infinite mass at absolute zero
   d. the pressure of the gas would rise to infinity at absolute zero

13. The "heat death" of the universe means that
   a. the universe will eventually become so hot that living organisms will not be able to survive
   b. the universe will be slowly destroyed as unstable atomic nuclei decay and release heat energy
   c. all forms of usable energy will ultimately become transformed into unusable heat energy
   d. the heat of friction caused by the gravitational collapse of the universe will convert all matter into heat energy

14. A candy bar contains 220 dietary calories. What is the energy content of the candy bar in kilocalories?
   a. 220 kcal
   b. 220,000 kcal
   c. 0.220 kcal
   d. 924 kcal

15. Which of the following is a heat engine operating in reverse?
   a. gasoline engine of an automobile
   b. jet airplane engine
   c. home air conditioner
   d. steam engine of a locomotive

16. A refrigerator ice maker converts water into ice cubes. During this process the entropy of the water
   a. increases
   b. decreases
   c. remains unchanged
   d. is destroyed

17. A rock with a volume of 6.0 cm^3 has a mass of 30.0 g. The density of the rock is
   a. 0.2 g/cm^3
   b. 5.0 g/cm^3
   c. 18.0 g/cm^3
   d. 24.0 g/cm^3

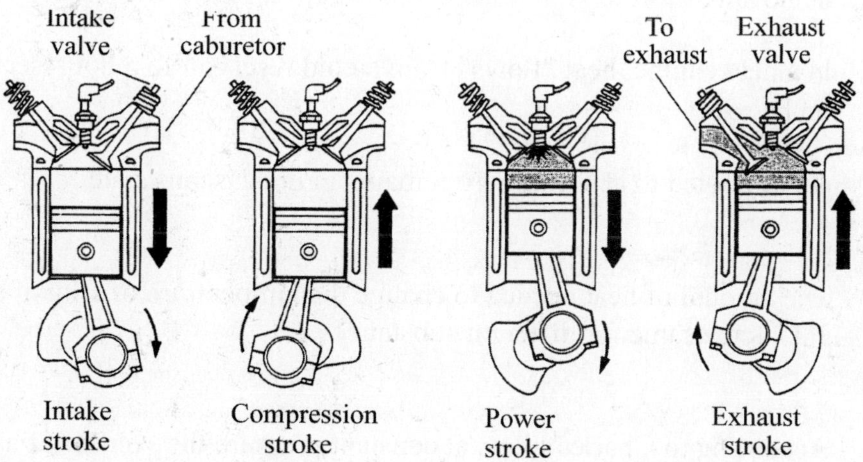

Refer to the above drawing to answer questions 18 through 20.

18. During which stroke is the volume of the fuel-air mixture reduced to 1/7 or 1/8 of its original volume?
   a. intake stroke
   b. compression stroke
   c. power stroke
   d. exhaust stroke

19. Immediately at the end of the compression stroke
   a. spent gases are vented out through the exhaust valve
   b. a mixture of air and gasoline vapor is drawn into the cylinder
   c. the piston is at its lowest point within the cylinder
   d. the spark plug is fired to ignite the fuel-air mixture

20. Heat energy is converted into work during the
   a. intake stroke
   b. compression stroke
   c. power stroke
   d. exhaust stroke

True or False

Decide whether each statement is true or false. If false, briefly state why it is false or correct the statement to make it true. See Chapter 1 or 2 for an example.

T 1. It is easier to convert other forms of energy into heat than to convert heat into other forms of energy.

F 2. In a heat engine, heat "flows" from a cold reservoir to a hot reservoir.

F 3. A gas cooled to absolute zero remains in the gaseous state.

F 4. The amount of heat needed to change the temperature of 1 kg of a substance by 1 °C is the same for different substances.

T 5. According to Charles's law, at constant pressure the volume of a gas is directly proportional to its absolute temperature.

T 6. If a gas at 257 K is heated to twice this temperature, or 514 K, the average kinetic energy of the gas molecules is doubled.

F 7. Large animals get rid of waste metabolic heat more efficiently than do smaller animals because of the larger animal's greater surface/mass ratio.

T 8. The absolute temperature of a gas is proportional to the average kinetic energy of its molecules.

F 9. The entropy of an isolated system tends to decrease with time.

T 10. It is impossible to transform all of the heat from a source into mechanical energy or work.

Fill in the Blank

1. On the F scale, water freezes at 32° F and boils at 212° F at sea level.
2. On the C scale, water freezes at 0° C and boils at 100° C at sea level.
3. The Joule is the SI unit of heat.
4. The Density of a substance is its mass per unit volume.
5. The Pascal is the SI unit of pressure.
6. A ______________ (2 words) is a device that converts heat into mechanical energy or work.
7. Water boils at 373 Kelvin.
8. Alcohol evaporates more rapidly than water because the attractive forces between its molecules are smaller.
9. In most electric generating plants, a large part of the energy lost as waste heat is due to the thermodynamic inefficiency of the ______________.

10. Compressing a gas causes its temperature to __increase__.

11. A heat engine converts part of the flow of the heat flowing from a hot reservoir to a cold one into ____________________.

12. The heat content of a given substance depends upon both its mass and its __temp__.

13. The __Color__ of an object hot enough to glow varies with its temperature.

14. In an aneroid barometer, the flexible ends of a sealed metal chamber are pushed in by a high __atmos__ pressure.

15. The ____________________ (2 words) of the universe will happen when all particles of matter have the same average kinetic energy and exist in a state of maximum entropy.

Matching

Match the term on the left with its definition on the right.

1._____ temperature

2._____ radiation

3._____ buoyant force

4._____ kilocalorie

5._____ thermostat

6._____ entropy

7._____ sublimation

8._____ convection

9._____ thermometer

10._____ conduction

11._____ thermodynamics

12._____ British Thermal Unit

a. heat transfer by electromagnetic waves
b. turns heating and cooling systems on and off
c. heat transfer by molecular collisions
d. the direct conversion of a solid to a vapor state
e. amount of heat needed to raise 1 lb of water by 1 °F
f. measure of average molecular KE
g. measures temperature
h. heat transport by the movement of a hot fluid
i. science of heat transformations
j. upward force that keeps an object afloat
k. measure of disorder
l. amount of heat needed to change the temperature of 1 kg of water by 1 °C

SOLVED PROBLEMS

Study the following solved example problems as they will provide insight into solving the problems listed at the end of Chapter 4 in *The Physical Universe.* Review the mathematics refresher in the text if you are unfamiliar with the basic mathematical operations presented in these examples.

Example 4-1

How many kJ of heat are needed to raise the temperature of 500 g of water from 60°C to 100°C?

Solution

It takes 4.2 kJ of heat to raise the temperature of 1 kg water 1°C. In this problem 500 g (0.500 kg) of water is raised a total of 40°C; therefore, the amount of heat needed to raise 500 g of water from 60°C to 100°C would be

$$(0.500\ \text{kg})(4.2\ \text{kJ/kg} \bullet {}^\circ\text{C})(40^\circ\text{C}) = 84\ \text{kJ}$$

Example 4-2

A certain quantity of hydrogen occupies a volume of 1000 cm^3 at 0°C (273 K) and ordinary atmospheric pressure. If the pressure is doubled but the temperature is held constant, what will the volume of the hydrogen be? If the temperature is increased to 172°C but the pressure is held constant, what will the volume of the hydrogen be?

Solution

The first part of the problem concerns the relationship between the volume and pressure of a gas sample at constant temperature; therefore, we must use Boyle's law to solve this part of the problem:

$$\frac{p_1}{p_2} = \frac{V_2}{V_1} \quad \text{(at constant temperature)}$$

The original pressure (normal atmospheric pressure) is 101 kPa. Doubled, this pressure (p_2) becomes 202 kPa. The initial volume is 1000 cm^3.

Substituting these quantities into the equation gives

$$\frac{101\ \text{kPa}}{202\ \text{kPa}} = \frac{V_2}{1000\ \text{cm}^3}$$

Solving for V_2 results in the new volume:

$$V_2 = \frac{(1000\ \text{cm}^3)(101\ \text{kPa})}{202\ \text{kPa}}$$

$$V_2 = \frac{101{,}000\ \text{cm}^3}{202}$$

$$V_2 = 500\ \text{cm}^3$$

Another way to think about this part of the problem is to remember that the volume of a gas sample is inversely proportional to its pressure at constant temperature. Double the pressure and the volume of the gas will be reduced to one-half of its original volume.

The second part of the problem concerns the relationship between the temperature and volume of a gas sample at constant pressure; therefore, we must use Charles's law to solve this part of the problem.

$$\frac{V_1}{V_2} = \frac{T_1}{T_2} \quad \text{(at constant pressure; temperatures on absolute scale)}$$

Substituting the known quantities into the equation gives

$$\frac{1{,}000\ \text{cm}^3}{V_2} = \frac{273\ \text{K}}{445\ \text{K}}$$

Note that the temperatures are absolute temperatures. The final temperature of 273 °C has been converted to 546 K:

Absolute temperature = celsius temperature + 273

Absolute temperature = 172°C + 273 = 445 K

Solving for V_2 results in the new volume:

$$V_2 = \frac{(1{,}000\ \text{cm}^3)(445\ \text{K})}{273\ \text{K}}$$

$$V_2 = \frac{445{,}000\ \text{cm}^3}{273}$$

$$V_2 = 1630\ \text{cm}^3$$

Example 4-3

An engine is designed to operate between 300°C and 100°C. Determine the engine's maximum efficiency.

Solution

The maximum efficiency of a heat engine is determined by the equation:

$$\text{Maximum efficiency} = 1 - \frac{T_{\text{cold}}}{T_{\text{hot}}}$$

The temperatures in the equation are absolute temperatures; therefore, the celsius temperatures given in the problem must be converted into absolute temperatures by adding 273 to each. We can now determine the maximum efficiency of the heat engine.

$$\text{Maximum efficiency} = 1 - \frac{373 \text{ K}}{573 \text{ K}}$$

Maximum efficiency = 1 - 0.651

Maximum efficiency = 0.349 which is 34.9 percent

WEB LINKS

Learn about the Ideal Gas law through a series of interactive experiments at:

http://jersey.uoregon.edu/vlab/Piston/index.html

ANSWER KEY

Multiple Choice

1. b 2. c 3. a 4. d 5. b 6. b 7. d 8. a 9. b 10. b 11. d 12. a 13. c 14. a 15. c 16. b 17. b 18. b 19. d 20. c

True or False

1. True
2. False. Heat "flows" from a hot reservoir to a cold reservoir in a heat engine.

3. False. It is impossible to cool a gas to absolute zero, or 0 K, because it is impossible to reach such a low temperature and because all known gases turn into liquids before absolute zero is reached.
4. False. The amount of heat needed to change the temperature of 1 kg of a substance by 1 °C is different for different substances.
5. True
6. True
7. False. Larger animals have a problem getting rid of metabolic heat because of their smaller surface/mass ratio.
8. True
9. False. The entropy of an isolated system cannot decrease and tends to increase with time.
10. True

Fill in the Blank

1. fahrenheit
2. celsius
3. joule
4. density
5. pascal
6. heat engine
7. 373
8. molecules
9. turbine
10. increase
11. mechanical energy
12. temperature
13. color
14. atmospheric
15. heat death

Matching

1. f 2. a 3. j 4. l 5. b 6. k 7. d 8. h 9. g 10. c 11. i 12. e

Chapter 5

ELECTRICITY AND MAGNETISM

OUTLINE

GOALS

1. Discuss what is meant by electric charge.
2. Describe the structure of an atom.
3. State Coulomb's law for electric force and compare it with Newton's law of gravity.
4. Account for the attraction between a charged object and an uncharged one.
5. Distinguish among conductors, semiconductors, and insulators.
6. Define ion and give several ways of producing ionization.
7. Define superconductivity and discuss it potential importance.
8. Describe electric current and potential difference (voltage) by analogy with the flow of water in a pipe.
9. Use Ohm's law to solve problems that involve the current in a circuit, the resistance of the circuit, and the voltage across the circuit.
10. Relate the power consumed by an electrical appliance to the current in it and the voltage across it.
11. Describe what is meant by a magnetic field and discuss how it can be pictured by field lines.
12. State the connection between electric charges and magnetic fields.
13. Explain how an electromagnet works.
14. Describe the force a magnetic field exerts on an electric current.
15. Discuss the operation of an electric motor.
16. Describe electromagnetic induction and explain how a generator makes use of it to produce an electric current.

CHAPTER SUMMARY

Electricity, the related phenomenon of magnetism, and the electrical nature of matter are discussed in Chapter 5. The nature of **electric charge** is presented, and the force between electric charges is described by **Coulomb's law**. The structure of the atom is introduced, and the three elementary particles found in atoms are identified by name, charge, and mass. **Electric current** is defined, and the concepts of **conductors** and **insulators** are presented. Electric current, voltage, and resistance are related by **Ohm's law**. Electric power is defined. The characteristics of magnets are discussed, and the relationship between electric current and magnetism is presented. The role of magnetic fields in the operation of electric motors, generators, transformers, maglev trains, and television picture tubes is explored.

CHAPTER OUTLINE

5-1. Positive and Negative Charge

A. Electric charge is a fundamental property of certain of the elementary particles of which all matter is composed.
B. There are two types of electric charge: **negative charge** and **positive charge**.
C. Like charges repel one another; unlike charges attract one another.
D. An object having equal amounts of positive and negative charges is said to be electrically **neutral**.

5-2. What Is Charge?

A. Every substance is composed of tiny bits of matter called **atoms**, which are composed of three kinds of **elementary particles**:
   1. The positively charged **proton**
   2. The negatively charged **electron**
   3. The electrically neutral **neutron**
B. The structure of an atom consists of a central **nucleus** of protons and neutrons with electrons moving about the nucleus.
C. An atom is electrically neutral since its number of protons (positive charges) is equal to its number of electrons (negative charges).
D. The **coulomb** (C) is the unit of electric charge.
E. The basic quantity of electric charge (e) is 1.6×10^{-19} C.

5-3. Coulomb's Law

A. **Coulomb's law** states that the force between two electrical charges is inversely proportional to the square of the distance between the charges and directly proportional to the product of the magnitude of the charges:

$$F = K\frac{Q_1Q_2}{R_2}$$

where F = force, K = a constant (9×10^9 N • m^2/C^2), Q = magnitude of charge, and R = distance between charges.

B. The coulomb is a very large unit. Even the most highly charged objects contain a small fraction of a coulomb of charge.

5-4. Force on an Uncharged Object

A. An electrically charged body attracts small uncharged objects such as dust particles and bits of paper.

B. The attractive force results from a separation of charge (positive and negative charges) induced in the uncharged object(s) by the charged body.

5-5. Matter in Bulk

A. Coulomb's law resembles the law of gravity; however, gravitational forces are always attractive, whereas electric forces may be attractive or repulsive.

B. Gravitational forces dominate on a cosmic scale; electric forces dominate on an atomic scale.

5-6. Conductors and Insulators

A. A **conductor** is a substance through which electric charge flows readily.

B. An **insulator** is a substance that strongly resists the flow of electric charge.

C. **Semiconductors** are substances whose electrical conductivity is between that of conductors and insulators.

D. **Transistors** are semiconductor devices whose conductivity can be changed at will.

E. The conduction of electricity through gases and solids involves the movement of charged atoms or molecules called **ions**.

1. A positive ion results when an atom or molecule loses electrons.
2. A negative ion results when an atom or molecule gains electrons.
3. The process of forming ions is called **ionization**.

5-7. Superconductivity

A. **Superconductivity** refers to the loss of all electrical resistance by certain materials at very low temperatures.

B. Superconductors allow the transmission of electric current without energy loss.

C. The use of superconductors was limited because they had to be cooled to extremely low temperatures; however, superconductors have been discovered that do not need the extreme cold of earlier ones.
D. The goal is to discover superconductors that operate at room temperatures.

5-8. The Ampere

A. **Electric current** is the rate of flow of charge from one place to another. Currents are by conventiion assumed to flow from the + terminal of a battery or other source to its – terminal.
B. A closed path that conducts an electric current is called a **circuit**.
C. The **ampere** (A) is the SI unit of electric charge and is equal to a flow of 1 coulomb per second:

$$1 \text{ A} = 1 \text{ C/s}$$

5-9. Potential Difference

A. **Potential difference**, or **voltage**, is the electrical potential energy per coulomb of charge.
   1. Potential difference is thought of as the work a unit charge can perform as the charge travels from one point to another.
   2. The **volt** (V) is the unit of potential difference and is equal to one joule per coulomb:

$$1 \text{ V} = 1 \text{ J/C}$$

B. The voltages of two or more batteries are added when the method of connection between the batteries is — terminal to + terminal.
C. Storage batteries can be **recharged** and used again.

5-10. Ohm's Law

A. **Ohm's law**, discovered by Georg Ohm (1787-1854), states that current in an electrical circuit is equal to voltage across the circuit divided by the resistance of the circuit:

$$I = \frac{V}{R}$$

where I = current, V = voltage, and R = resistance.

B. **Resistance** is a measure of opposition to the flow of charge.

1. The **ohm** (Ω) is the unit of resistance and is equal to 1 volt per ampere.

$$1\ \Omega = 1\ \text{V/A}$$

2. The resistance of a metal wire depends on its:
   a. Composition: Some materials offer more resistance than others.
   b. Length: The longer the wire, the greater the resistance.
   c. Cross-sectional area: The greater the wire's cross-sectional area, the less the resistance.
   d. Temperature: The higher the temperature, the more the resistance.

C. Ohm's law holds only for solid metallic conductors.

5-11. Electric Power

A. The **power** of an electric current is the rate at which it does work and is equal to the product of the current and the voltage of a circuit:

$$P = IV$$

where P = power, I = current, and V = voltage.

B. The unit of electric power is the watt.

C. The commercial unit of electric energy is the **kilowatthour** (kWh).

5-12. Magnets

A. Every magnet has a **north pole** and a **south pole**.

B. Like magnetic poles repel one another; unlike poles attract.

C. Magnetic poles always come in pairs. There is no such thing as a single north pole or south pole.

D. All substances have magnetic properties, though generally to a very slight extent.

E. In a magnet, all of the atoms are aligned with their N poles in the same direction.

5-13. Magnetic Field

A. A **force field** is a region of altered space around a mass, an electric charge, or a magnet that exerts a force on appropriate objects in that region.

B. The force field around a magnet is called a **magnetic field**.

C. **Field lines** are imaginary lines in a field of force; they help us visualize a magnetic field.

5-14. Oersted's Experiment

A. Hans Christian Oersted discovered in 1820 that an electric current near a compass causes the compass needle to be deflected.
B. Oersted's experiment showed that every electric current has a magnetic field surrounding it.
C. According to the **right-hand rule**, the electron current in a wire and the magnetic field it generates are perpendicular to each other.
D. All magnetic fields originate from moving electric charges.
    1. A magnetic field appears only when relative motion is present between an electric charge and an observer.
    2. Electric and magnetic fields are different aspects of a single electromagnetic field.

5-15. Electromagnets

A. The magnetic field strength of a wire coil carrying an electric current increases in direct proportion to the number of turns of the coil.
B. An **electromagnet** consists of an iron core placed inside a wire coil.

5-16. Magnetic Force on a Current

A. A magnetic field exerts a sideways push on an electric current with the maximum push occurring when the current is perpendicular to the magnetic field.
B. Currents exert magnetic forces on each other. The forces are attractive when parallel currents are in the same direction and are repulsive when the parallel currents are in opposite directions.

5-17. Electric Motors

A. An electric motor uses the sideways push of a magnetic field to turn a current-carrying wire loop.
B. To prevent the loop from coming to rest when it is perpendicular to the magnetic field, direct current electric motors use a **commutator** to change the direction of the current in the loop.
C. Alternating current electric motors do not use commutators since the alternating current automatically changes the direction of the current in the wire loop.

5-18. Electromagnetic Induction

A. Michael Faraday discovered that an **induced current** is produced in a wire moved across a magnetic field.
   1. The direction of the induced current can be reversed by reversing the motion of the wire or reversing the field direction.
   2. The strength of the current depends on the strength of the magnetic field and the speed of the wire's motion.

B. The effect of producing an induced current is known as **electromagnetic induction**.
   1. **Alternating current** (ac) is current that flows in a back-and-forth manner; household current changes direction 120 times each second (60 Hz).
   2. **Direct current** (dc) flows in one direction.

C. The ac generator (or **alternator**) produces an ac current and can be modified to produce dc current by
   1. Use of a commutator.
   2. Use of a **rectifier** which permits current to pass through it in only one direction.

5-19. Transformers

A. A **transformer** is a device composed of two unconnected coils, usually wrapped around a soft iron core, that can increase or decrease the voltage of ac current.

B. The induced voltage in the secondary coil can be higher, lower, or the same as that of the primary coil depending on the ratio of turns between the primary and secondary coils.
   1. If the number of turns of both coils is equal, the induced voltage is also equal.
   2. If the secondary coil has more turns, the induced voltage is higher.
   3. If the secondary coil has fewer turns, the induced voltage is lower.

C. Electric power ($P=IV$) is always the same in both coils; thus the relationship between the number of turns, voltage, and current of a transformer is represented by the formula:

$$\frac{N_1}{N_2} = \frac{V_1}{V_2} = \frac{I_2}{I_1} \quad (N = \text{number of turns})$$

KEY TERMS AND CONCEPTS

The questions in this section will help you review the key terms and concepts from Chapter 5.

Multiple Choice

Circle the best answer for each of the following questions.

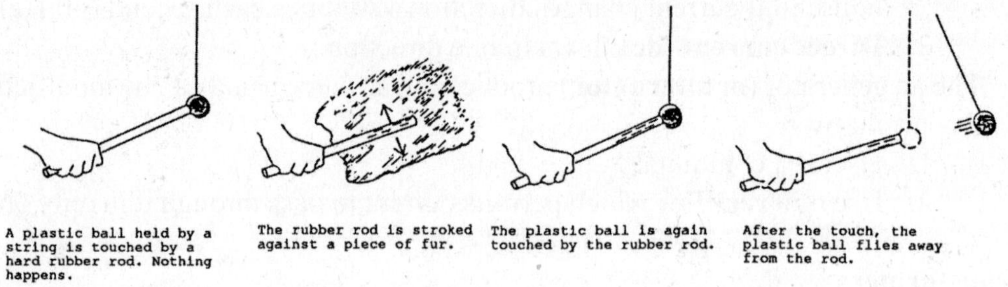

Refer to the above drawing to answer questions 1 through 3.

1. When the rubber rod is stroked against the piece of fur
   a. electrons are transferred from the fur to the rubber rod
   b. electrons are transferred from the rubber rod to the fur
   c. the fur becomes negatively charged
   d. the rubber rod becomes magnetized

2. When the plastic ball is touched by the rubber rod
   a. some of the charge is transferred from the rubber rod to the plastic ball
   b. charge flows from the plastic ball to the rubber rod
   c. the charge in the rubber rod is neutralized
   d. no charge is transferred between the rubber rod and plastic ball

3. The plastic ball flies away from the rubber rod because
   a. its electric field has been neutralized
   b. it has acquired an electric charge opposite that of the rubber rod
   c. some of the charge flows from the rod to the plastic ball and like charges repel
   d. the rod's magnetic field pushes the plastic ball away

4. The elementary particle that has no electric charge is the
   a. proton
   b. electron
   c. atomic nucleus
   d. neutron

5. The ampere is the unit of
   a. resistance
   b. electric current
   c. potential difference
   d. charge

6. If you divide voltage by resistance, the answer will be in units of
   a. coulombs
   b. ohms
   c. amperes
   d. watts

7. Ohm's law holds only for
   a. metallic conductors
   b. gaseous conductors
   c. liquid conductors
   d. transistors

8. The unit of electric power is the
   a. volt
   b. ohm
   c. ampere
   d. watt

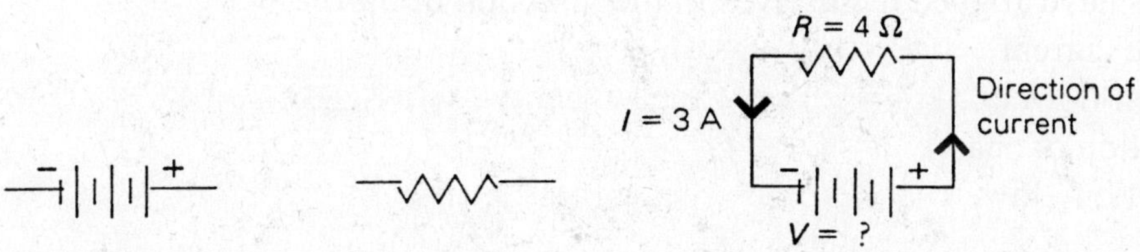

A B C

Refer to the above drawings to answer questions 9 through 11.

9. Drawing A is the symbol for a
   a. magnet
   b. battery
   c. compass
   d. transformer

10. Drawing B is the symbol for
   a. electric power
   b. resistance
   c. wattage
   d. voltage

11. In Drawing C, the voltage is
   a. 1.5 V
   b. 0.75 V
   c. 12 V
   d. 7 V

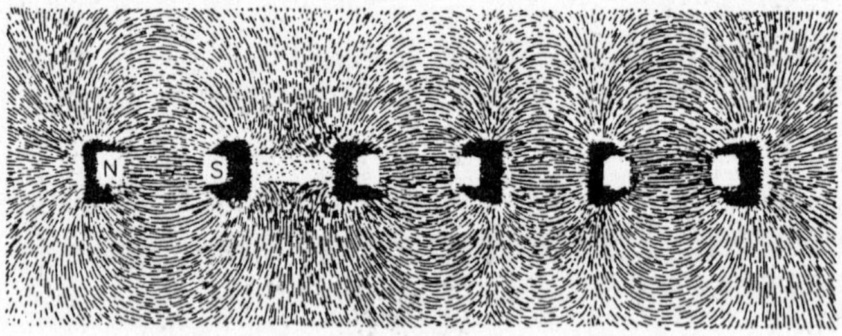

A B C

Refer to the above drawing to answer questions 12 and 13. The drawing represents patterns formed by iron filings sprinkled on a card held over three bar magnets (A, B, C).

12. The filings have aligned themselves in the direction of the
   a. electric current
   b. electron flow
   c. separation of charge
   d. magnetic field

13. In the drawing, the poles for magnets B and C are unlabeled. Beginning with magnet B, the correct order for the missing poles is
   a. N, S, S, N
   b. S, N, N, S
   c. S, N, S, N
   d. N, S, N, S

14. Oersted's experiment demonstrated that
    a. a compass needle points toward the earth's north magnetic pole
    b. a magnetic field surrounds every electric current
    c. field lines surround every magnet
    d. an electric charge at rest has magnetic properties

15. A _____ circuit results when the connecting wires of an electric circuit accidently touch each other or are joined by a stray conductor.
    a. live
    b. random
    c. short
    d. hot

True or False

Decide whether each statement is true or false. If false, either briefly state why it is false or correct the statement to make it true. See Chapter 1 or 2 for an example.

_________ 1. Rubbing a glass rod with silk creates positive electric charges within the glass rod, giving it an overall positive charge.

_________ 2. The force between two charged objects depends upon the distance between the objects and the magnitude of each charge.

_________ 3. The electrons in a metallic conductor carry the current.

_________ 4. Electric energy is useful partly because it is easily transformed into other kinds of energy.

_________ 5. Ohm's law applies to metallic and liquid conductors, but not to gaseous conductors.

_________ 6. Cutting a magnet in half results in one-half having a single north pole and no south pole, and the other half having a single south pole and no north pole.

_________ 7. Heating or hammering a magnet destroys the magnetic fields of its atoms or molecules.

T ________ 8. A stationary electric charge has no magnetic properties.

T ________ 9. In a transformer, the electrical power is the same in both coils independent of the number of turns in each coil.

F ________ 10. The force a magnetic field exerts on an electric current disappears when the current is perpendicular to the magnetic field.

Fill in the Blank

1. An electrically neutral atom has equal numbers of protons and ___electrons___.

2. Neutrons and protons are found in the ___nucleus___ of an atom.

3. ___Electrons___ are elementary particles found in atoms and have a negative electrical charge.

4. The ___watt___ is the basic unit of electric charge.

5. A ___coulomb___ is a substance through which electric charge can flow.

6. In an ___insulator___, charge can flow only with great difficulty.

7. ___Ions___ are atoms or molecules that have lost or gained one or more electrons.

8. ___Grounding___ is the method of removing the charge from an object.

9. ___Superconductivity___ is the loss of all electrical resistance by certain materials at extremely low temperatures.

10. A flow of charge from one place to another is an electric ___current___.

11. The ___voltage___, or potential difference, between two points is the work needed to take a charge of 1 C from one of the points to the other.

12. The ___volt___ is the unit of potential difference.

13. ___Resistance___ is the property of a conductor that opposes the flow of charge passing through it.

14. The unit of electrical resistance is the ___ohm___.

15. The ___Power___ of an electric current is the rate at which it does work.

16. Electric power is the product of current and ___voltage___.

16. When the resistors in an electric circuit are joined end-to-end with the same current flowing through all of them, the resistors are in ___series___; when the resistors are arranged so that the total current is split among them, they are in ___parallel___.

18. In a simple direct-current electric motor, the ___commutater___ reverses the current periodically

19. When electric currents are transmitted, some of the electrical energy is lost as ___heat___.

20. A ___Fuse___ or circuit breaker interrupts a power line whenever an unsafe amount of current passes through it.

Matching

Match the name of the person on the left with the description on the right.

1.______ Michael Faraday

2.______ Christian Oersted

3.______ Benjamin Franklin

4.______ Charles Coulomb

5.______ Georg Ohm

6.______ Andre Ampere

7.______ Thales of Miletus

8.______ Alessandro Volta

9.______ Alex Muller and George Bednorz

10.______ Kamerlingh Onnes

a. the unit of electric current is named for him

b. first to discover a material superconducting above 23 K

c. discovered that every electric current generates a magnetic field

d. made the first recorded studies of electricity

e. described charges as either positive or negative

f. the unit of potential difference is named for him

g. discovered the principle of the generator

h. showed how the force between charges varied with their distance apart and the magnitude of each charge

i. discovered the phenomenon called superconductivity

k. the unit of resistance is named for him

SOLVED PROBLEMS

Study the following solved example problems as they will provide insight into solving the problems listed at the end of Chapter 5 in *The Physical Universe*. Review the mathematics refresher in the Appendix of the Study Guide if you are unfamiliar with the basic mathematical operations presented in these examples.

Example 5-1

If a 60-W light bulb is connected to a 120-V power line, how much current flows through it? What is the resistance of the bulb? How much power does the bulb consume?

Solution

To solve the first part of the problem, note that the problem gives the power of the light bulb in watts (60 W) and the voltage of the circuit (120 V). From this information it is possible to

determine the current (I) that flows through the light bulb by using the equation for electric power:

$$P = IV$$

Solving for I the equation becomes

$$I = \frac{P}{V}$$

Substituting the information given in the problem and solving for I we get

$$I = \frac{P}{V} = \frac{60\ \text{W}}{120\ \text{V}} = 0.5\ \text{A}$$

To determine the resistance of the bulb, use Ohm's law:

$$I = \frac{V}{R}$$

Solving for R the equation becomes

$$R = \frac{V}{I}$$

Substituting for V and I and solving for R we get

$$R = \frac{120 \text{ V}}{0.5 \text{ A}} = 240 \ \Omega$$

Since the watt is the unit of electrical power, a 60-W bulb uses 60 W of power. The power of the bulb can also be calculated by using the equation for power:

$$P = IV = (0.5 \text{ A})(120 \text{ V}) = 60 \text{ W}$$

Example 5-2

A telephone pole transformer rated at a maximum power of 10 kW is used to couple a 7200-V transmission line to a 240-V circuit. What is the ratio of turns in the transformer? What is the maximum current in the 240-V circuit?

Solution

The first part of the problem asks for the ratio of turns in the transformer but does not provide the number of turns in the primary and secondary coils; however, the ratio of the primary turns to the secondary turns is equal to the ratio of the primary voltage and the secondary voltage. These quantities are given in the problem:

$$\frac{N_1}{N_2} = \frac{V_1}{V_2} = \frac{7200 \text{ V}}{240 \text{ V}} = 30$$

The second part of the problem asks for the maximum current in the 240-V circuit. Current is equal to power divided by voltage. Since the power of a transformer is the same for both coils, the calculation is

$$I = \frac{P}{V} = \frac{10 \text{ kW}}{240 \text{ V}} = \frac{10{,}000 \text{ W}}{240 \text{ V}} = 41.7 \text{ A}$$

WEB LINKS

Investigate Ohm's law at this interactive web site

http://www.walter-fendt.de/ph11e/ohmslaw.htm

Determine the relationship between voltage, amperage, and resistance at these interactive web sites

http://jersey.uoregon.edu/vlab/Voltage/

http://webphysics.ph.msstate.edu/jc/library/18-2/simulation.html

Operate a virtual multimeter to measure current, voltage, and resistance at this web site

http://www.phy.ntnu.edu.tw/java/electronics/multimeter.html

Investigate the behavior of a DC electric motor at this interactive web site

http://webphysics.ph.msstate.edu/jc/library/20-3/simulation.html

ANSWER KEY

Multiple Choice

1. a 2. a 3. c 4. d 5. b 6. c 7. a 8. d 9. b 10. b 11. c 12. d 13. c 14. b 15. c

True or False

1. False. The glass rod, prior to rubbing with silk, was electrically neutral, meaning it had equal amounts of positive and negative charges. Rubbing with silk removed some of the rod's negative charges (electrons), giving the rod an overall positive charge.
2. True
3. False. The electrons in a metallic conductor do not carry the current; the moving electrons <u>are</u> the current.
4. True
5. False. Ohm's law applies only to metallic conductors and not to gaseous or liquid conductors.
6. False. Each half has both a north pole and a south pole; magnetic poles always comes in pairs.
7. False. The magnetic properties of the magnet's atoms or molecules are not destroyed by heating or hammering the magnet. Instead, the atoms and molecules become randomly arranged, which cancels out their collective magnetic effect.
8. True
9. True
10. False. The force a magnetic field exerts on an electric current is perpendicular to the magnetic field.

Fill in the Blank

1. electrons
2. nucleus
3. electrons
4. coulomb
5. conductor
6. insulator
7. ions
8. grounding
9. superconductivity
10. current

11. voltage
12. volt
13. resistance
14. ohm
15. power
16. voltage
17. series, parallel
18. commutator
19. heat
20. fuse

Matching

1. g 2. c 3. e 4. h 5. k 6. a 7. d 8. f 9. b 10. i

Chapter 6
WAVES

OUTLINE

GOALS

1. State what a wave is and give examples of different kinds of waves.
2. Distinguish between transverse and longitudinal waves.
3. Use the formula $v = \lambda f$ to relate the frequency and wavelength of a wave to its speed.
4. Discuss the nature of sound.
5. State what the doppler effect is and explain its origin.
6. Describe the various ways in which musical instruments produce sound.
7. Discuss the nature of electromagnetic waves and give examples of different types of such waves.
8. Distinguish between amplitude and frequency modulation.
9. Explain how a radar works.
10. Describe what is meant by a light ray.
11. Explain how a mirror produces an image.
12. Explain what is meant by the refraction and the reflection of waves.
13. Explain how refraction makes a body of water seem shallower than it actually is.
14. Explain what is meant by internal reflection.
15. Define lens and distinguish between converging and diverging lenses.
16. Describe the differences between farsightedness, near sightedness, and astigmatism.
17. Account for the dispersion of white light into a spectrum when it is refracted.
18. Discuss the origin of rainbows.
19. Explain why the sky is blue.
20. Distinguish between constructive and destructive interference.
21. Explain why thin films of soap or oil are brightly colored.
22. Describe the diffraction of waves at the edge of an obstacle.
23. Describe the effect of diffraction on the sharpness of the images produced by optical instruments.

CHAPTER SUMMARY

Chapter 6 discusses **waves**. Waves fall into two important categories: **mechanical waves** and **electromagnetic waves**. The descriptive characteristics of waves are discussed, and the behavior of waves as they interact with various mediums, physical obstacles, and other waves is presented. The wave nature of sound and light is emphasized, and musical sounds, optical phenomena, and radio communication are explained in terms of wave behavior and propagation.

CHAPTER OUTLINE

6-1. Water Waves

A. A **wave** is a periodic disturbance that moves away from a source and carries energy as it goes.

B. Two important categories of waves are:

1. **Mechanical waves,** which travel only through matter and involve the motion of particles of the matter they pass through. Sound waves are an example.
2. **Electromagnetic waves,** which consist of varying electric and magnetic fields and can travel through a vacuum and through matter; they do not involve the motion of particles of the matter they pass through. Light, radio waves, and X-rays are examples.

6-2. Transverse and Longitudinal Waves

A. **Transverse waves** are mechanical waves in which the particles of the matter through which they pass move perpendicular to the wave direction; transverse waves can travel only through solids.

B. **Longitudinal waves** are mechanical waves in which the particles of the matter through which they pass move parallel to the wave direction in a series of **compressions** and **rarefactions**; longitudinal waves can travel through fluids as well as solids.

C. Water waves are a combination of both transverse and longitudinal waves.

6-3. Describing Waves

A. **Wavelength** (λ) is the distance from crest to crest (or trough to trough) of a wave.

B. **Frequency** (f) is the number of crests that pass a given point each second; the unit of frequency (cycles per second) is the **hertz** (Hz).

C. **Speed** (v) is the rate at which each crest moves; wave speed is equal to wavelength times frequency:

$$v = \lambda f$$

D. **Period** (T) is the time needed for a wave to pass a given point.
E. **Amplitude** (A) is the maximum displacement from a normal position of the particles of the medium through which a wave passes.

6-4. Standing Waves

A. **Standing waves** occur when reflected waves interact with forward-moving waves in such a way that some points in the medium have amplitudes twice that of the normal amplitude and at other points the amplitude is zero. Such waves appear to be stationary or standing still.

6-5. Sound

A. Sound waves are longitudinal waves.
   1. Speed of sound is about 343 m/s (767 mi/h) in sea-level air at ordinary temperatures.
   2. Sound travels faster in liquids and solids than in gases.
B. The **decibel** (dB) is the unit of sound intensity.
C. Sounds with frequencies below about 20 Hz are called **infrasound**; those above about 20,000 Hz are called **ultrasound**.
D. The human ear is most sensitive to sound frequencies between 3000 and 4000 Hz.
E. Ultrasound used in echo-sounding is called **sonar**.

6-6. Doppler Effect

A. The **doppler effect** is the apparent change in frequency of a wave due to the relative motion of the listener and the source of the sound.
   1. As the relative motion reduces the distance between the source and the observer, the frequency, or pitch, of the sound becomes higher.
   2. As the relative motion increases the distance between the source and the observer, the pitch becomes lower.
B. The doppler effect also occurs in light waves and is used by astronomers to calculate the speed at which stars are approaching or receding.

6-7. Musical Sounds

A. Musical sounds are produced by vibrating objects such as strings, vocal cords, membranes in drums, and air columns in wind instruments.
   1. **Fundamental tone** is the tone produced when an object vibrates as a whole; this is always the lowest frequency.
   2. **Overtones** are higher frequencies that are produced when an object vibrates in segments; they add richness and quality, or timbre, to the fundamental tone.

B. **Resonance** is the ability of an object to be set in vibration by a source whose frequency is equal to one of its natural frequencies of vibration.
C. The fundamental frequencies in ordinary human speech are mostly below 1000 Hz, averaging about 145 Hz in men and about 230 Hz in women.

6-8. Electromagnetic Waves

A. Maxwell proposed that a magnetic field is associated with a changing electric field.
B. Electromagnetic (em) waves consist of linked electric and magnetic fields traveling at the speed of light.
C. The electric and magnetic fields in an em wave form closed loops that spread out from their source with the speed of light.
D. The electric and magnetic fields in an em wave are perpendicular to each other and to the direction of the wave.

6-9. Types of EM Waves

A. In 1887, the German physicist Heinrich Hertz demonstrated the existence of em waves.
B. Electromagnetic waves can carry information as well as energy.
C. Radio communication uses **amplitude modulation** (AM) or **frequency modulation** (FM).
    1. Amplitude modulation is the modulation of a radio wave by varying its amplitude.
    2. Frequency modulation is the modulation of a radio wave by varying its frequency.
D. **Radar** (radio detection and ranging) is a device that uses ultra high frequency em waves to detect distant objects such as ships and airplanes.
E. The electromagnetic **spectrum** is the range of frequencies of em waves.

6-10. Light "Rays"

A. Light does not always travel a straight path since it can be reflected and refracted.
B. Light appears to travel in a straight path in a uniform medium and, although light actually consists of waves, it is useful to represent light's straight line motion as lines called **rays.**

6-11. Reflection

A. **Reflection** is the change in direction of a wave when it strikes an obstacle.
B. The **image** in a mirror appears to originate from behind the mirror.
C. In a mirror image, left and right are interchanged because front and back have been reversed by the reflection.

6-12. Refraction

A. **Refraction** is the change in direction of a train of waves when they enter a medium in which their speed changes.
B. Refraction occurs when waves cross a boundary at a slanting angle; if waves approach a boundary at right angles, no refraction occurs.
C. Light is refracted when it goes from one medium into another medium in which the speed of light is different.
D. The amount of deflection when light is refracted depends on the speeds of light in the two mediums.
E. The **index of refraction** is the ratio between the speed of light in free space and its speed in a medium.
F. The **internal reflection** of light occurs when the angle through which a light ray passing from one medium to a less optically dense medium is refracted by more than 90°.

6-13. Lenses

A. A **lens** is a piece of glass or other transparent material shaped to produce an image by refracting light that comes from an object.
B. A **converging** lens is thicker in the middle than at its rim and brings parallel light rays to a single focal point, called the **real focal point** because the light rays pass through it, at a distance called the **focal length** of the lens.
C. A **diverging** lens is thinner in the middle than at its rim and spreads out parallel light rays so that they seem to come from a point behind the lens called the **virtual focal point** because the light rays do not pass through it but appear to do so.

6-14. The Eye

A. The human eye operates much like a camera and consists of the following structures:
   1. A transparent outer membrane called the **cornea.**
   2. A **lens**, which focuses incoming light onto the **retina**, which contains light-sensitive receptors.
   3. The **optic nerve,** which carries nerve impulses from the retina to the brain.
   4. **Ciliary muscles,** which change the shape, and thus focal length, of the lens.
   5. The colored **iris,** which controls the amount of light entering the **pupil** or opening of the iris.

B. There are two common vision defects:
   1. **Farsightedness** occurs when the eyeball is too short, focusing the object behind the retina and making it difficult to focus on nearby objects.
   2. **Nearsightedness** occurs when the eyeball is too long, focusing the object in front of the retina and making it difficult to focus on distant objects.

C. **Astigmatism** occurs when the cornea or lens has different curvatures in different planes.

6-15. Color

A. **White light** is a mixture of light waves of different frequencies.
B. Each frequency of light produces the visual sensation of a particular color.
C. **Dispersion** is the separation of a beam of white light into its various colors or frequencies; rainbows are caused by the dispersion of sunlight by water droplets.
D. An object's color depends on the kind of light that falls on the object and on the nature of its surface.

6-16. Interference

A. **Interference** refers to the adding together of two or more waves of the same kind that pass by the same point at the same time.
   1. In **constructive interference**, the original waves are in step and combine to give a wave of greater amplitude.
   2. In **destructive interference**, the original waves are out of step and combine to give a wave of smaller amplitude.
B. When light of only one color (one wavelength) strikes a thin film, the film appears dark where the light waves reflected from its upper and lower surfaces undergo destructive interference; the film appears bright where constructive interference takes place.
C. When white light strikes a thin film, the reflected waves of only one color will be in step at a particular place while waves of other colors will not; this interference results in a series of brilliant colors.
D. Standing waves are formed by interference.

6-17. Diffraction

A. **Diffraction** is the ability of waves to bend around the edge of an obstacle.
B. Diffracted waves spread out as though they originated at the corner of the obstacle and are weaker than the direct waves.
C. Because of diffraction, the images of microscopes and telescopes are blurred at high magnification.
D. The larger the diameter of a lens or mirror used in an optical instrument, the less significant the diffraction and the sharper the image.
E. The resolving power of a telescope depends upon the wavelength of the light that enters it divided by the diameter of the lens or mirror; the smaller the resolving power, the sharper the image.

KEY TERMS AND CONCEPTS

The questions in this section will help you review the key terms and concepts from Chapter 6.

Multiple Choice

Circle the best answer for each of the following questions.

1. Radio waves and x-rays are examples of
   a. electromagnetic waves
   b. mechanical waves
   c. pressure waves
   d. longitudinal waves

2. Which one of the following waves cannot travel through a vacuum?
   a. sound waves
   b. radio waves
   c. light waves
   d. x-rays

3. Which one of the following wave quantities is not related to the others?
   a. speed
   b. frequency
   c. amplitude
   d. wavelength

4. Human ears are most sensitive to sounds whose frequencies are between
   a. 10,000 and 20,000 Hz
   b. 3000 and 4000 Hz
   c. 5000 and 8000 Hz
   d. 20 and 1500 Hz

5. Two waves having the same frequency and equal amplitude A meet exactly in step at the same point. The resulting amplitude is
   a. 0.5 A
   b. 2 A
   c. unchanged
   d. zero

6. A violin string is bowed and at the same time is touched lightly in the middle, causing the string to vibrate in two sections. The two vibrating sections produce sound frequencies called
   a. fundamental tones
   b. undertones
   c. overtones
   d. supertones

7. If the violin string in Problem 6 was not touched and allowed to vibrate as a whole, its frequency relative to the frequencies of the two vibrating sections would be
   a. the same
   b. lower
   c. higher
   d. higher or lower depending upon the energy used to make it vibrate

8. A stationary observer hears the whistle of an approaching train. To the observer, the frequency of the sound produced by the whistle appears to be
   a. lowest before the train reaches the observer
   b. lowest just as the train passes the observer
   c. the same before the train reaches the observer as after the train passes the observer
   d. higher before the train reaches the observer and lower after the train passes the observer

9. The portion of the eye that contains the light-sensitive receptors is the
   a. lens
   b. cornea
   c. pupil
   d. retina

10. A narrow beam of white light is passed through a prism and is separated into beams of light of various colors. This is an example of
    a. interference
    b. dispersion
    c. total internal reflection
    d. resonance

True or False

Decide whether each statement is true or false. If false, either briefly state why it is false or correct the statement to make it true. See Chapter 1 or 2 for an example.

F __________ 1. In a transverse wave, the particles of the medium move parallel to the wave direction.

_________ 2. Sound waves travel faster in air than in liquids or solids.

_________ 3. Electromagnetic waves differ from mechanical waves in their ability to travel through a vacuum as well as matter.

_________ 4. Hearing the noise of a car horn around the corner of a building is an example of reflection.

_________ 5. The size and shape of a musical instrument is largely responsible for its natural frequencies of vibration or resonance.

_________ 6. In 1864, James Clerk Maxwell proposed that an accelerated electric charge generates combined electrical and magnetic disturbances called electromagnetic waves.

_________ 7. The smaller the *f*-number of a camera lens, the smaller the lens opening.

_________ 8. A person suffering from nearsightedness can see distant objects clearly, but cannot focus on nearby objects.

_________ 9. The sun appears a brilliant red at sunset because the atmospheric molecules and dust particles scatter the blue light more than the red.

_________ 10. The speed of light in a medium such as air or glass is slightly different for different frequencies.

Fill in the Blank

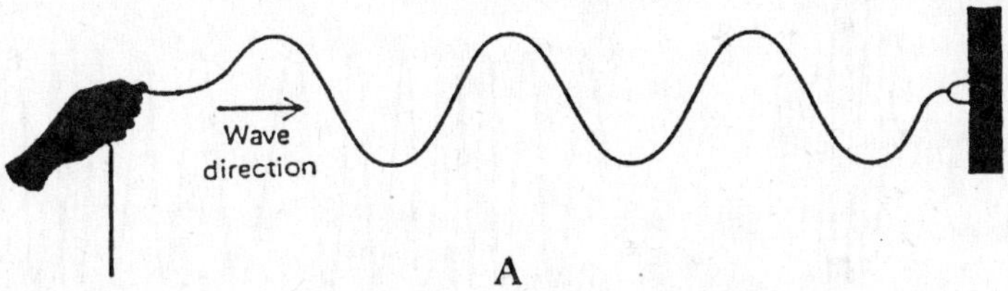

A

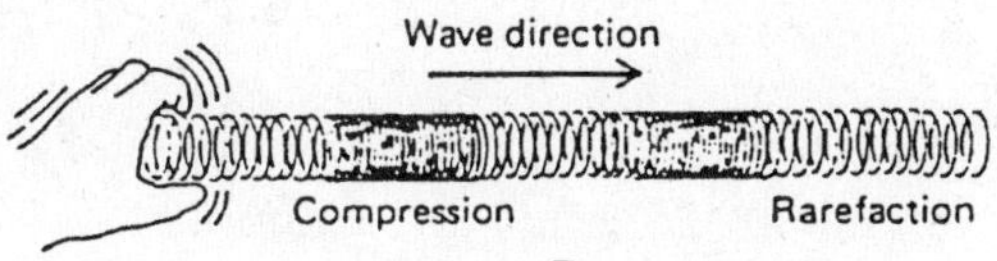

B

Refer to the above drawing to answer questions 1 through 4.

1. The type of waves generated in (A) are ____________ waves.

2. In (A), the particles of the rope move ____________ to the direction of the waves.

3. The type of waves generated in (B) are ____________ waves.

4. In (B), the particles of the spring move ____________ to the direction of the wave.

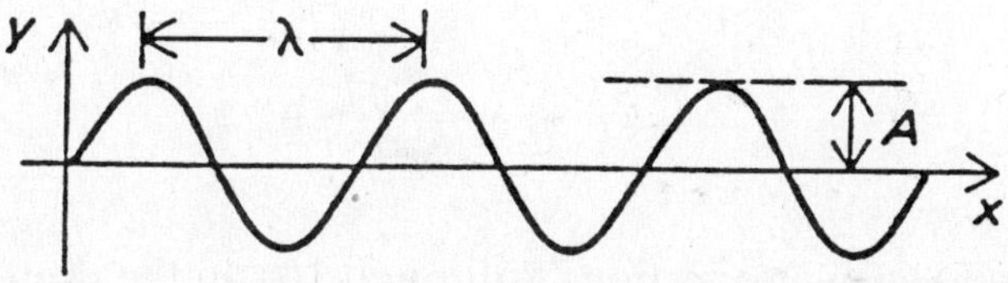

5. In the above drawing, the Greek letter lambda (λ) represents the wave's ____________ and *A* represents the wave's ____________.

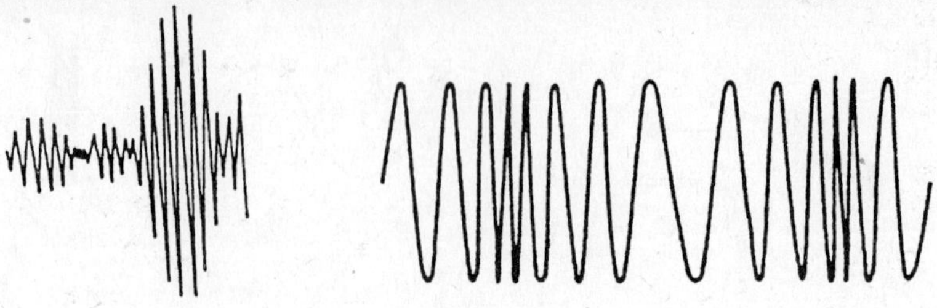

A B

Refer to the above drawing to answer questions 6 and 7; (A) represents an AM radio signal and (B) represents an FM radio signal.

6. In (A), the ___Frequency___ of the radio waves remains constant and the ___amplitud___ of the waves is modulated.

7. In (B), the ___amplitude___ of the radio waves remains constant and the ___Frequency___ of the waves is modulated.

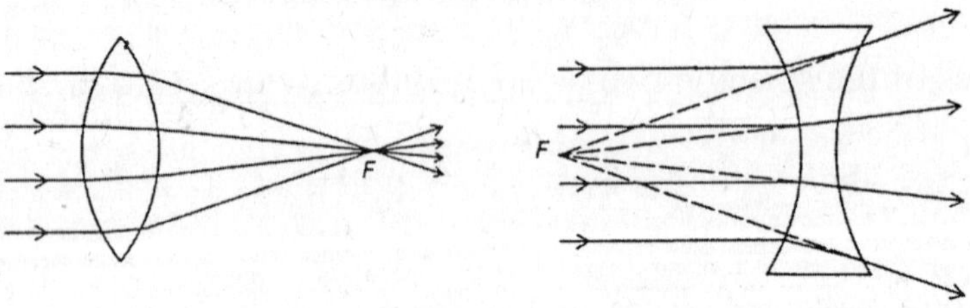

A B

Refer to the above drawing to answer questions 8 through 10. In the drawing, two different types of lenses are represented.

8. The lens illustrated in (A) is a ___converging___ lens; the lens illustrated in (B) is a ___diverging___ lens.

9. In (A), the point labeled *F*, where the light rays converge, is called the ___Focal___ point.

10. In both (A) and (B), the distance between *F* and the lens is called the ___Focal length___ (2 words) of the lens.

Matching

1._____ wavelength

2._____ reflection

3._____ ultrasound

4._____ period

5._____ hertz

6._____ wave speed

7._____ decibel

8._____ dispersion

9._____ infrasound

10._____ refraction

a. unit of frequency
b. time needed for a wave to pass a given point
c. unit of sound intensity
d. change in a wave's direction as it enters a different medium
e. distance between adjacent wave crests or troughs
f. sound frequencies below 20 Hz
g. equal to wavelength times frequency
h. separation of white light into its individual colors or frequencies
i. sound frequencies above 20,000 Hz
j. change in a wave's direction when it strikes an obstacle

SOLVED PROBLEMS

Study the following solved example problems as they will provide insight into solving the problems listed at the end of Chapter 6 in *The Physical Universe*. Review the mathematics refresher in the Appendix of the Study Guide if you are unfamiliar with the basic mathematical operations presented in these examples.

Example 6-1

A piano string vibrates 4186 times per second. How many vibrations does it make while its sound travels 1.0 m? (Hint: the speed of sound in air is 343 m/s)

Solution

A piano string vibrating at a constant frequency generates a wavetrain composed of waves of equal wavelength. Since each vibration of the string creates one wave, the number of vibrations of the string, as its sound travels 1.0 m, would be equal to the number of waves of wavelength λ that travel the 1.0-m distance. The problem's solution consists of two steps:

1. Calculate the value for wavelength

2. Determine the number of waves by dividing the 1.0-m distance by the value for the wavelength

To calculate the wavelength, use the equation for wave speed and solve for λ.

The speed of sound in air is 343 m/s. The frequency is 1044 vibrations per second or 1044 Hz. Substituting these quantities into the equation for wavelength we get

$$\lambda = \frac{v}{f} = \frac{343 \text{ m/s}}{4186 \text{ Hz}} = 0.082 \text{ m}$$

The number of waves is

$$\frac{d}{\lambda} = \frac{1.0 \text{ m}}{0.082 \text{ m}} = 12$$

The number of vibrations = number of waves = 12.

Example 6-2

A typical radio controlled toy truck operates at a frequency of 75 MHz. What is the corresponding wavelength?

Solution

To solve the problem, use the equation for wavelength. Remember that a radio wave travels with the speed of light, or 3.00 x 10^8 m/s, and that 1 MHz = 10^6 Hz.

$$\lambda = \frac{v}{f} = \frac{3.00 \times 10^8 \text{ m/s}}{75 \text{ MHz}} = \frac{3.00 \times 10^8 \text{ m/s}}{7.5 \times 10^7 \text{ Hz}} = 4 \times 10^0 \text{ m} = 4 \text{ m}$$

WEB LINKS

Investigate the properties of a converging lens at

http://webphysics.ph.msstate.edu/javamirror/ipmj/java/clens/index.html

Investigate the properties of a diverging lens at

http://webphysics.ph.msstate.edu/javamirror/ipmj/java/dlens/index.html

Investigate reflection and refraction at these web sites

http://www.phy.ntnu.edu.tw/~hwang/light/flashLight.html

http://webphysics.ph.msstate.edu/jc/library/22-2b/simulation.html

ANSWER KEY

Multiple Choice

1. a 2. a 3. c 4. b 5. b 6. c 7. b 8. d 9. d 10. b

True or False

1. False. The particles move perpendicular to the wave direction in a transverse wave.
2. False. Sound waves travel faster in liquids and solids.
3. True
4. False. Hearing the noise of a car horn around the corner of a building is an example of diffraction.
5. True
6. True
7. False. The smaller the *f*-number, the larger the lens opening.
8. False. A nearsighted person can see nearby objects clearly but has difficulty focusing on distant objects.
9. True
10. True

Fill in the Blank

1. transverse
2. perpendicular
3. longitudinal
4. parallel
5. wavelength, amplitude
6. frequency, amplitude
7. amplitude, frequency
8. converging, diverging
9. focal
10. focal length

Matching

1. e 2. j 3. i 4. b 5. a 6. g 7. c 8. h 9. f 10. d

Chapter 7

THE NUCLEUS

OUTLINE

GOALS

1. Discuss the connection between the Rutherford experiment and the modern picture of atomic structure.
2. Distinguish between nucleon and nuclide and between atomic number and mass number.
3. State in what ways the isotopes of an element are similar and in what ways they are different.
4. Describe the various kinds of radioactive decay and explain why each occurs.
5. Define half-life.
6. Discuss the sources and hazards of the ionizing radiation we are exposed to in daily life.
7. Explain the significance of the binding energy of a nucleus.
8. Draw a graph of binding energy per nucleon versus mass number and indicate on it the location of the most stable nucleus and the range of mass numbers in which fusion and fission can occur.
9. Discuss nuclear fission and the conditions needed for a chain reaction to occur.
10. Describe how a nuclear reactor works.
11. Discuss the good and bad features of nuclear reactors as energy sources.
12. Describe nuclear fusion and identify the conditions needed for a successful fusion reactor.
13. Compare a particle with its antiparticle.
14. Describe the process of annihilation and pair production.
15. List the four fundamental interactions and identify the aspects of the universe that each governs.
16. Distinguish between leptons and hadrons and discuss the quark model of hadrons.

CHAPTER SUMMARY

The structure of atomic nuclei and the importance of nuclear reactions and transformations are the main topics in Chapter 7. Rutherford's discovery of the nucleus is summarized and related to modern atomic theory. The occurrence of elemental **isotopes** is explained and the types of **radioactive decay** are described. The concept of nuclear **binding energy** is presented and its relationship to the processes of **nuclear fission** and **nuclear fusion** is developed. The operation of nuclear power plants and the future of nuclear power plants and the future of nuclear energy is discussed. **Leptons** and **hadrons** are introduced as categories of elementary particles and examples of each are presented. **Quarks** are defined and their role in the structures of protons and neutrons is explored.

CHAPTER OUTLINE

7-1. Rutherford Model of the Atom

A. In 1898, British physicist J. J. Thompson described atoms as positively charged lumps of matter with electrons embedded in them.

B. In 1911, an experiment suggested by British physicist Ernest Rutherford shows that alpha particles striking a thin metal foil are deflected by the strong electric fields of the metal atom's nuclei.

C. Rutherford's experiment resulted in the following conclusions:

1. An atom's positive charge and nearly all of its mass are concentrated in its nucleus.
2. The atom's electrons, because of their small mass, have little effect on the alpha particles.
3. An atom is mostly empty space.

7-2. Nuclear Structure

A. The nucleus of ordinary hydrogen is a single positively charged **proton**; other nuclei contain electrically neutral **neutrons** as well as protons.

B. The number of protons in an atom is the **atomic number**; the number of protons equals the number of electrons.

C. **Isotopes** are atoms of the same element that differ in the number of neutrons in their nuclei.

D. A nucleus with a particular composition is called a **nuclide** and is represented by

$$^{A}_{Z}X$$

where X = chemical symbol, Z = atomic number, and A = **mass number** or the number of protons and neutrons in the nucleus.

E. A **nucleon** is a neutron or proton; the mass number of a nucleus is the number of nucleons (protons and neutrons) it contains.

7-3. Radioactive Decay

A. In 1896, Henri Becquerel discovered that uranium gives off a penetrating radiation, a property called **radioactivity**.
B. Soon after Becquerel's discovery, Pierre and Marie Curie discovered two more radioactive elements: polonium and radium.
C. **Radioactive decay** occurs when a nucleus emits particles or high frequency em waves.
D. There are five kinds of radioactive decay.
    1. Alpha decay, in which an **alpha particle** ($^{4}_{2}He$) is emitted from a large, unstable nucleus.
    2. Gamma decay, in which very high frequency em waves, or **gamma rays**, are emitted by nuclei having excess energy.
    3. Beta decay, in which an electron is emitted from a nucleus having an unstable neutron to proton ratio; a nuclear neutron changes to a proton and an electron.
    4. Positron emission, a type of beta decay in which a **positron** (a positively charged electron) is emitted from a nucleus having an unstable proton to neutron ratio; a nuclear proton changes to a neutron and a positron.
    5. **Electron capture**, in which the capture of an electron by a nuclear proton changes the proton to a neutron in a nucleus having a high proton to neutron ratio.

7-4. Half-Life

A. The **half-life** of a radionuclide (radioactive nuclide) is the time needed for half of an original sample to decay.
B. Half-lives vary widely among radionuclides, ranging from less than a millionth of a second to billions of years.

7-5. Radiation Hazards

A. All ionizing radiation is harmful to living tissue.
B. The SI unit of radiation dosage is the **sievert** (Sv); 1 Sv is the amount of radiation having the same biological effects as those produced when 1 kg of body tissue absorbs 1 J of x-rays or gamma rays.
C. Some examples of radioactive sources are:
    1. radon gas
    2. radionuclides in rocks and soil
    3. radionuclides that occur naturally in the body
    4. nuclear reactors

5. medical x-ray machines
6. cosmic rays

7-6. Units of Mass and Energy

A. The **atomic mass unit** (u) is the standard unit of atomic mass:

$$1 \text{ atomic mass unit} = 1 \text{ u} = 1.66 \times 10^{-27} \text{ kg}$$

B. The **electronvolt** (eV) is the energy unit used in atomic physics:

$$1 \text{ electronvolt} = 1 \text{ eV} = 1.60 \times 10^{-19} \text{ J}$$

C. The **megaelectronvolt** (MeV) is equal to 1 million eV:

$$1 \text{ megaelectronvolt} = 1 \text{ MeV} = 10^{6} \text{ eV} = 1.60 \times 10^{-13} \text{ J}$$

D. The energy equivalent of a rest mass of 1 u is 931 MeV.

7-7. Binding Energy

A. All atoms have <u>less</u> mass than the combined masses of the particles of which they are composed.
   1. When a nucleus is formed, a certain amount of energy is given off due to the action of the forces holding the neutrons and protons together.
   2. The energy comes from the mass of the particles that join together.

B. The energy equivalent of the missing mass of a nucleus is called the **binding energy**; the greater the binding energy of a nucleus, the more the energy needed to break it apart.

7-8. Binding Energy per Nucleon

A. For a given nucleus, the **binding energy per nucleon** is found by dividing the total binding energy of the nucleus by the number of nucleons (protons and neutrons) it contains; the greater the binding energy per nucleon, the more stable the nucleus.

B. **Nuclear fission** is the splitting of a heavy nucleus into two medium-sized ones, each one having <u>less</u> binding energy per nucleon than the original nucleus.

C. **Nuclear fusion** is the joining of two light nuclei to give a single nucleus of medium size having <u>less</u> binding energy per nucleon than the two original nuclei.

D. Binding energy makes stable heavier nuclei possible (beyond hydrogen) which in turn accounts for the various elements and forms of matter found in the physical universe.

7-9. Nuclear Fission

A. In 1939, uranium-235 was discovered to undergo nuclear fission when struck by a neutron.
   1. A nucleus of U-235 absorbs the neutron to become U-236.
   2. The U-236 nucleus is unstable and splits into two smaller nuclei.

B. A **chain reaction** is a series of fission reactions spreading through a mass of an unstable radionuclide such as uranium.
   1. When a nucleus undergoes fission, two or three neutrons are released and can cause other nuclei to split and begin a chain reaction.
   2. The first chain reaction was demonstrated by the Italian physicist Enrico Fermi in Chicago in 1942.

C. In a nuclear reactor, fission occurs at a controlled rate.

D. In an atomic bomb, uncontrolled fission produces an explosion.

7-10. How a Reactor Works

A. A nuclear power plant transforms nuclear energy into electricity.
   1. Controlled nuclear fission within the reactor produces heat.
   2. The heat is removed by a liquid or gas coolant.
   3. The hot coolant converts water into steam.
   4. The steam is fed to a turbine that powers an electric generator.

B. The chain reaction within a nuclear reactor is controlled by a **moderator** which slows down neutrons.

C. A water-moderated reactor uses the hydrogen in water (H_2O) as the moderator; such reactors use **enriched** uranium as a fuel.

7-11. Plutonium

A. When nonfissionable U-238 captures a fast neutron, it eventually forms the fissionable nuclide of plutonium, Pu-239, which can support a chain reaction.

B. Plutonium is a **transuranium element**, meaning that it has an atomic number greater than the 92 of uranium.

C. The fissionable plutonium produced in a uranium-fueled reactor can be used as a fuel or in nuclear weapons.

7-12. A Nuclear World?

A. Nuclear energy generates about 21 percent of the electricity produced in the United States.

B. Questions of safety, costs, and nuclear waste disposal have halted construction of nuclear reactors in the United States.

C. Accidents at the Three Mile Island reactor in Pennsylvania in 1979, and at Ukraine's Chernobyl reactor in 1986, have caused some nations to slow down or abandon their nuclear energy programs; other countries still heavily rely on nuclear power.

7-13. Nuclear Fusion

A. Nuclear fusion produces tremendous quantities of energy and has the potential of becoming the ultimate source of energy on earth.

B. There are three requirements for a fusion reaction:
   1. A high temperature of 100 million °C or more.
   2. A high concentration of nuclei to assure frequent collisions.
   3. A confinement time long enough to allow reacting nuclei to give off more energy than is used in the reactor's operation.

C. Current fusion research involves the fusion of deuterium nuclei ($^{2}_{1}H$) and tritium nuclei ($^{3}_{1}H$) to form helium nuclei ($^{4}_{2}He$), neutrons, and energy.

D. There are three approaches to make fusion energy practical by overcoming the problems of temperature, density of reacting nuclei, and confinement time.
   1. One approach uses strong magnetic fields to contain the reacting material.
   2. The pellet method uses laser beams or beams of charged particles (electrons or protons) to heat and compress tiny deuterium-tritium pellets.
   3. The **Z-pinch** method sends a strong electric current to a cage of tungsten wires surrounding a fuel pellet. The resulting magnetic forces collapse the tungsten cage compressing the fuel pellet and heating it to over 10 million °C.

7-14. Antiparticles

A. An **elementary particle** cannot be separated into other particles.

B. Nucleons (protons and neutrons) are composed of smaller particles called **quarks**; however, nucleons are considered to be elementary particles because their quarks are so tightly bound that the nucleons cannot be split.

C. An **antiparticle** has the same mass and general behavior as its corresponding elementary particle, but has a charge of opposite sign and differs in certain other respects.

D. When an antiparticle and its corresponding elementary particle come together, they undergo **annihilation,** with their masses turning entirely into energy.

E. In the process of **pair production**, a particle-antiparticle pair materializes from energy.

7-15. Fundamental Interactions

A. Four fundamental interactions give rise to all physical processes in the universe; in order of decreasing strength these are:

1. The **strong interaction**, which holds protons and neutrons together to form atomic nuclei.
2. The **electromagnetic interaction**, which gives rise to electric and magnetic forces between charged particles.
3. The **weak interaction**, which, by causing beta decay, helps determine the compositions of atomic nuclei.
4. The **gravitational interaction**, which is responsible for the attractive force one mass exerts on another.

7-16. Leptons and Hadrons

A. Elementary particles can be categorized into two groups based on their response to the strong interaction.
   1. **Leptons**, which are not affected by the strong interaction, have no internal structure.
      a. Electrons are leptons.
      b. **Neutrinos** are leptons that have no charge and very little mass.
   2. **Hadrons**, which are affected by the strong interaction, are composed of quarks; protons and neutrons are hadrons.

B. Quarks have two unusual properties:
   1. They have fractional charges.
   2. They do not exist independently outside of hadrons.

KEY TERMS AND CONCEPTS

The questions in this section will help you review the key terms and concepts from Chapter 7.

Short Answer

For questions 1 through 6, write the name of the fundamental interaction(s) that accounts for each structure in the blank to the right of the drawing. Question 1 has been completed as an example.

1. Molecule

1. <u>electromagnetic interaction</u>

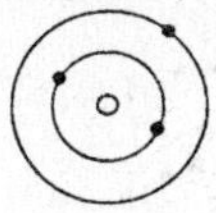

2. Atom 2. electromagnetic

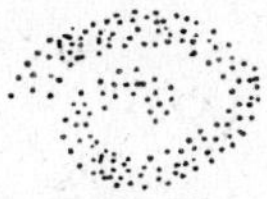

3. Galaxy of stars 3. gravitational

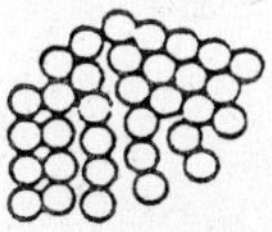

4. Solid, liquid 4. electromagnetic

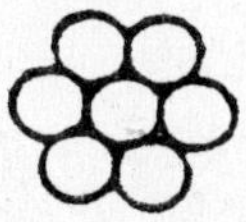

5. Atomic Nucleus 5. strong & weak

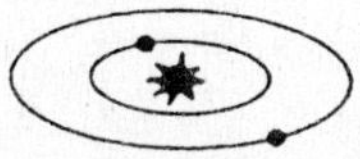

6. Planet, star 6. gravitational

Multiple Choice

Circle the best answer for each of the following questions.

1. The nuclei of all atoms other than ^1_1H contain
   a. electrons and protons
   b. neutrons and protons
   c. neutrons and electrons
   d. protons only

2. Rutherford's experimental procedure, which revealed the structure of the atom, involved
   a. aiming a beam of x-rays at a block of lead
   b. aiming neutrons at a mass of fissionable uranium
   c. aiming a narrow beam of alpha particles at a thin gold foil
   d. causing collisions of fast-moving electrons with protons and neutrons

3. Rutherford's experiment revealed that
   a. atoms are mostly empty space
   b. all atoms have the same atomic number
   c. quarks are the building blocks of protons and neutrons
   d. atoms are electrically neutral

4. At room temperature and normal atmospheric pressure, most elements are
   a. unstable
   b. gases
   c. liquids
   d. solids

5. The mass number of a nucleus is the number of
   a. protons in the nucleus
   b. neutrons in the nucleus
   c. protons and neutrons in the nucleus
   d. quarks in the nucleus

6. An electron that has a positive charge is called a
   a. neutrino
   b. positron
   c. hadron
   d. nuclide

7. The most important single source of the radiation dosage received by an average person in the United States is
   a. cosmic rays
   b. radon gas
   c. x-ray machines
   d. nuclear reactors

8. One atomic mass unit is approximately equal to the mass of a(n)
   a. electron
   b. atom of hydrogen
   c. alpha particle
   d. neutrino

9. The energy needed to break a nucleus apart is called its
   a. energy of fission
   b. nuclear energy
   c. binding energy
   d. separation energy

10. No new nuclear reactors have been ordered in the United States since 1979 because
   a. the United States is energy self-sufficient and no new reactors are needed
   b. the United States has run out of uranium
   c. the United States is waiting until nuclear fusion becomes practical before building new reactors
   d. of public concerns over safety of nuclear reactors and unfavorable economic conditions

11. Currently, the best method of disposing of nuclear wastes seems to be
   a. incineration
   b. underground storage
   c. ocean dumping
   d. chemical neutralization

12. Which one of the following is not a requirement for a fusion reactor?
   a. a moderator of graphite or water
   b. extremely high temperatures to initiate fusion
   c. sufficient confinement time for a net energy output
   d. a high concentration of fuel nuclei

13. During the process of pair production
   a. matter is converted into energy
   b. a particle is converted into its antiparticle counterpart
   c. energy is converted into matter
   d. an electron decays to form a pair of quarks

14. The fundamental interaction that has yet to be included in a grand unified theory is the
   a. strong interaction
   b. electromagnetic interaction
   c. weak interaction
   d. gravitational interaction

15. An example of a lepton is the
   a. electron
   b. quark
   c. proton
   d. neutron

True or False

Decide whether each statement is true or false. If false, either briefly state why it is false or correct the statement to make it true. See Chapter 1 or 2 for an example.

T 1. The atomic number of an element, because it determines how many electrons its atoms have and how they are arranged, governs the physical and chemical behavior of the element.

F 2. All atoms of a given element have nuclei with the same number of protons and the same number of neutrons.

T 3. An atom always has less mass than the sum of the masses of its protons, neutrons, and electrons.

F 4. When a neutron strikes the nucleus of a fissionable uranium-235 atom, it is the impact of the collision that splits the nucleus into two pieces.

F 5. The pellet method of nuclear fusion uses an intense magnetic field to compress a pellet of deuterium-tritium fuel and generate the heat needed for fusion.

T 6. The electromagnetic interaction is responsible for the structures of atoms, molecules, liquids, and solids.

F 7. Leptons are elementary particles that are affected by the strong interaction.

Fill in the Blank

1. In the Rutherford model of the atom, the positive charge is concentrated in a central ____________________.

2. The ____atomic____ number is the number of protons in an atom's nucleus.

3. The ____isotopes____ of an element have atoms whose nuclei differ in their numbers of neutrons.

4. The ____mass____ number of a nucleus is the number of nucleons it contains.

5. A ____nuclide____ is an atom whose nucleus has particular atomic and mass numbers.

6. ____Alpha____ particles are helium nuclei.

7. ____Beta____ particles are electrons.

8. ____Gamma____ rays are high-energy frequency electromagnetic waves.

9. A ____positron____ is a positively charged electron.

10. The ____half-life____ of a radioactive nuclide is the period of time required for one-half of the initial amount of the nuclide to decay.

11. The SI unit of radiation dosage is the ____sievert____.

12. Nuclear ____Fission____ occurs when a heavy nucleus splits into lighter nuclei.

13. Nuclear ____Fusion____ occurs when two nuclei combine.

14. A ____moderator____ is a substance use in a nuclear reactor to slow down fast neutrons.

15. An ____elementary____ particle cannot be separated into other particles.
16. Protons and neutrons consist of smaller particles called ____quarks____.

17. The ____________________ of an elementary particle has the same mass and general behavior but its electric charge is opposite in sign.

18. The ____________________ interaction helps determine the composition of atomic nuclei.

19. The ____________________ interaction holds protons and neutrons together in atomic nuclei.

20. ____________________ are leptons that have no charge and little or no mass.

Matching

Match the name of the person on the left with the description on the right. Write the letter of the correct description in the blank beside the name.

1.______ Henri Becquerel

2.______ Paul A. M. Dirac

3.______ J. J. Thompson

4.______ Enrico Fermi

5.______ Ernest Rutherford

6.______ Pierre and Marie Curie

7.______ Steven Weinberg and Abdus Salam

a. demonstrated the first nuclear chain reaction
b. discovered the radioactive elements polonium and radium
c. discovered the radioactive nature of uranium
d. independently linked the weak and electromagnetic interactions
e. said atoms were positively charged lumps of matter containing embedded electrons
f. established the structure of the atom
g. predicted the existence of positrons

SOLVED PROBLEMS

Study the following solved example problems as they will provide insight into solving the problems listed at the end of Chapter 7 in *The Physical Universe*. Review the mathematics refresher in the Appendix of the Study Guide if you are unfamiliar with the basic mathematical operations presented in these examples.

Example 7-1
Oxygen-22 (atomic number 8) undergoes two successive negative beta decays. Find the atomic number, mass number, and chemical name of the resulting nucleus.

Solution

Negative beta decay involves the emission of an electron by a nuclear neutron, which changes the neutron to a proton. If O-22 undergoes two successive negative beta decays, it will end up with two additional nuclear protons; therefore, the atomic number of the resulting nucleus is 10. The mass number of the new nucleus remains unchanged at 22 because, even though the number of protons has increased by two and the number of neutrons has decreased by two, the total number of nuclear protons and neutrons is the same. The chemical name of the resulting nucleus is <u>neon</u> since neon is the element having the atomic number 10.

Example 7-2

Polonium-209 decays to lead-205 through alpha decay. If an alpha particle with a mass of 6.6 x 10^{-27} kg leaves the polonium nucleus at a speed of 1.6 x 10^{7} m/s, what is the kinetic energy of the alpha particle in MeV?

Solution

The equation for kinetic energy (from Chapter 3) is

$$\text{KE} = \tfrac{1}{2}mv^2$$

The mass of the alpha particle is 6.6 x 10^{-27} kg; the particle's speed is 1.6 x 10^{7} m/s. Substituting these values into the equation for KE we get

$$\text{KE} = \tfrac{1}{2}(6.6 \times 10^{-27}\ \text{kg})(1.6 \times 10^{7}\ \text{m/s})^2$$

$$= \tfrac{1}{2}(6.6 \times 10^{-27}\ \text{kg})(1.6 \times 10^{7}\ \text{m/s})(1.6 \times 10^{7}\ \text{m/s})$$

$$= \tfrac{1}{2}(6.6 \times 10^{-27}\ \text{kg})(2.6 \times 10^{14}\ \text{m}^2/\text{s}^2)$$

$$\text{KE} = 8.6 \times 10^{-13}\ \text{kg/m}^2/\text{s}^2 = 8.6 \times 10^{-13}\ \text{J}$$

The problem asks for the electron's KE in MeV, not J. Since 1 MeV = 1.6 x 10^{-13} J, then

$$\text{KE (in MeV)} = \frac{8.6 \times 10^{-13}\ \text{J}}{1.6 \times 10^{-13}\ \text{J}} = 5.37\ \text{MeV}$$

Example 7-3

The mass of ${}^{7}_{3}\text{Li}$ is 7.0160 u. Find its binding energy and binding energy per nucleon.

Solution

The binding energy of a ${}^{7}_{3}\text{Li}$ nucleus is the energy equivalent of the missing mass representing the difference in the mass of the ${}^{7}_{3}\text{Li}$ and the sum of the masses of its protons, neutrons, and electrons. To calculate the binding energy of ${}^{7}_{3}\text{Li}$, the amount of the missing mass must first be determined and then converted into its energy equivalent.

The $^{7}_{3}Li$ atom may be regarded as consisting of three hydrogen atoms (each containing a proton and an electron) and four neutrons. The mass of a hydrogen atom is 1.0078 u and the mass of a neutron is 1.0087 u. The combined mass of three hydrogen atoms and four neutrons is

$$3\ m_H + 4\ m_n = (3 \times 1.0078\ u) + (4 \times 1.0087\ u) = 7.0582\ u$$

The difference in mass between this combined mass and the actual mass of the $^{4}_{2}He$ atom is

$$7.0582\ u - 7.0160\ u = 0.0422\ u$$

The energy equivalent of the atomic mass unit (u) is 931 MeV; therefore, the binding energy of $^{4}_{2}He$ is

$$(0.0422\ u)(931\ MeV/u) = 39.3\ MeV$$

The binding energy per nucleon for a given nucleus is found by dividing its total binding energy by the number of nucleons (protons and neutrons) it contains. Thus the binding energy per nucleon for $^{7}_{3}Li$ is

$$\frac{39.3\ MeV}{7} = 5.6\ MeV$$

WEB LINKS

Take a virtual tour of an atom at

http://library.thinkquest.org/11771/english/hi/physics/vrml2.shtml

See an interactive demonstration of radioactive decay at

http://home.earthlink.net/~mmc1919/halflife.html

Watch virtual beta decay at

http://www.colorado.edu/physics/2000/applets/iso.html

Operate a virtual nuclear power plant and prevent a possible meltdown at

http://www.ida.liu.se/~her/npp/demo.html

ANSWER KEY

Short Answer

1. electromagnetic
2. electromagnetic
3. gravitational
4. electromagnetic
5. strong and weak
6. gravitational

Multiple Choice

1. b 2. c 3. a 4. d 5. c 6. b 7. b 8. b 9. c 10. d 11. b 12. a 13. c 14. d 15. a

True or False

1. True
2 False. All atoms of a given element have nuclei with the same number of protons but not necessarily the same number of neutrons.
3. True
4. False. It is not the impact of a neutron that causes the fission of an uranium-235 nucleus; instead the nucleus absorbs the neutron to become uranium-236, which is so unstable that it undergoes fission almost immediately.
5. False. The pellet method of nuclear fusion uses laser beams or beams of charged particles such as electrons and protons to heat and compress tiny deuterium-tritium pellets.
6. True.
7. False. Leptons are unaffected by the strong interaction.

Fill in the Blank

1. nucleus
2. atomic
3. isotopes
4. mass
5. nuclide
6. alpha
7. beta
8. gamma
9. positron
10. half-life
11. sievert
12. fission
13. fusion
14. moderator
15. elementary
16. quarks
17. antiparticle
18. weak
19. strong
20. neutrinos

Matching

1. c 2. g 3. e 4. a 5. f 6. b 7. d

Chapter 8

THE ATOM

OUTLINE

GOALS

1. Describe the photoelectric effect and discuss why the wave theory of light cannot account for it.
2. Explain how the quantum theory of light accounts for the photoelectric effect.
3. Compare the quantum and wave theories of light and discuss why both are needed.
4. Describe x-rays and interpret their production in terms of the quantum theory of light.
5. Discuss what is meant by the matter wave of a moving particle.
6. State the uncertainty principle and interpret it in terms of matter waves.
7. Distinguish between emission and absorption spectra.
8. Give the basic ideas of the Bohr model of the atom and show how they follow from the wave nature of moving electrons.
9. Define quantum number, energy level, ground state, and excited state.
10. Explain how a laser works.
11. List the three characteristic properties of laser light.
12. Compare quantum mechanics and newtonian mechanics.
13. Describe what is meant by the probability cloud of an atomic electron.
14. State the exclusion principle.

CHAPTER SUMMARY

Chapter 8 introduces the science of **quantum mechanics** and discusses how modern quantum theory has enhanced our understanding of puzzling phenomena such as the **photoelectric effect**, the wave-particle nature of light and of moving objects, and the behavior of atomic electrons. Heisenberg's **uncertainty principle** is stated, and newtonian mechanics is shown to be an approximate version of quantum mechanics. The **Bohr model of the atom** is discussed and compared with the modern quantum theory of the atom. The four quantum numbers needed to specify the physical state of an atomic electron are discussed, and Pauli's **exclusion principle** is stated.

CHAPTER OUTLINE

8-1. Photoelectric Effect

A. The **photoelectric effect** is the emission of electrons from a metal surface when light shines on it.
B. The discovery of the photoelectric effect could not be explained by the electromagnetic theory of light.
C. Albert Einstein developed the **quantum theory of light** in 1905.

8-2. Photons

A. Einstein's quantum theory of light was based on a hypothesis suggested by the German physicist Max Planck in 1900.
   1. Planck stated that the light emitted by a hot object is given off in discrete units or **quanta**.
   2. The higher the frequency of the light, the greater the energy per quantum.
   3. All the quanta associated with a particular frequency of light have the same energy. The equation is

$$E = hf$$

where E = energy, h = **Planck's constant** (6.63×10^{-34} J • s), and f = frequency.
B. Einstein expanded Planck's hypothesis by proposing that light could travel through space as quanta of energy called photons. These photons, if of sufficient energy, could dislodge electrons from a metal surface causing the photoelectric effect.
C. Einstein's equation for the photoelectric effect is

$$hf = \text{KE} + w$$

where hf = energy of a photon whose frequency is f, KE = kinetic energy of the emitted electron, and w = energy needed to pull the electron from the metal.

D. Although photons have no mass and travel with the speed of light, they have most of the other properties of particles.

8-3. What Is Light?

A. Light exhibits either wave characteristics or particle (photon) characteristics, but never both at the same time.

B. The wave theory of light and the quantum theory of light are both needed to explain the nature of light and therefore complement each other.

8-4. X-rays

A. Wilhelm Roentgen accidentally discovered **x-rays** in 1895.

B. In 1912, Max von Laue showed that x-rays are extremely high frequency em waves.

C. X-rays are produced by high energy electrons that are stopped suddenly; the electron KE is transformed into photon energy.

8-5. De Broglie Waves

A. In 1924, the French physicist Louis de Broglie proposed that moving objects behave like waves; these are called **matter waves**.

B. The **de Broglie wavelength** of a particle of mass m and speed v is

$$\lambda = \frac{h}{mv}$$

where λ = de Broglie wavelength, h = Planck's constant, and mv = momentum of the particle.

C. Matter waves are significant only on an atomic scale.

D. A moving body exhibits wave properties in certain situations and exhibits particle properties in other situations.

8-6. Waves of What?

A. The quantity that varies in a matter wave is called the **wave function** (ψ).

B. The square of the wave function (ψ^2) is called the **probability density**. For a given object, the greater the probability density at a certain time and place, the greater the likelihood of finding the object there at that time.

C. The de Broglie waves of a moving object are in the form of a group, or packet, of waves that travel with the same speed as the object.

8-7. Uncertainty Principle

A. The **uncertainty principle** states that it is impossible to know both the exact position and momentum of a particle at the same time.
B. The discoverer of the uncertainty principle was Werner Heisenberg.
C. The position and motion of any object at a given time can only be expressed as probabilities.

8-8. Atomic Spectra

A. A gas whose electrons have absorbed energy is said to be **excited**.
B. A **spectroscope** is an instrument that disperses the light emitted by an excited gas into the different frequencies the light contains.
C. An **emission spectrum** consists of the various frequencies of light given off by an excited substance.
D. A **continuous spectrum** consists of all frequencies of light given off by an excited substance.
E. An **absorption spectrum** consists of the various frequencies absorbed by a substance when white light is passed through it.
F. The frequencies in the spectrum of an element fall into sets called **spectral series**.

8-9. The Bohr Model

A. The Niels Bohr model of the atom, proposed in 1913, suggested that an electron in an atom possesses a specific energy level that is dependent on the orbit it is in. An electron in the innermost orbit has the least energy.
B. Electron orbits are identified by a **quantum number** *n*, and each orbit corresponds to a specific **energy level** of the atom.
    1. Electrons cannot possess energies between specific energy levels or orbits.
    2. An electron can be raised to a higher energy level by absorbing a photon or, by emitting a photon, fall to a lower energy level.
    3. When an electron "jumps" from one orbit (energy level) to another, the difference in energy between the two orbits is hf, where h is the frequency of the emitted or absorbed light.
C. An atom having the lowest possible energy is in its **ground state**; an atom that has absorbed energy is in an **excited state**.

8-10. Electron Waves and Orbits

A. An electron can circle a nucleus only in orbits that contain a whole number of de Broglie wavelengths.
B. The quantum number *n* of an orbit is the number of electron waves that fit into the orbit.

8-11. The Laser

A. A **laser** is a device that produces an intense beam of single-frequency, **coherent** light from the cooperative radiation of excited atoms.
B. The word laser comes from light amplification by stimulated emission of radiation.
C. Lasers use materials whose atoms have **metastable states**, which are excited states with relatively long lifetimes.
   1. Ruby lasers use xenon-filled flash lamps to excite chromium ions in ruby rods.
   2. Helium-neon lasers use an electric discharge to bring the atoms of the gas mixture to metastable levels.
D. The metastable atoms, as they return to their ground states, create photons all of the same frequency and all of whose waves are coherent or exactly in step.

8-12. Quantum Mechanics

A. The theory of **quantum mechanics** was developed by Erwin Schrödinger, Werner Heisenberg, and others during the mid-1920s.
B. According to quantum mechanics, the position and momentum of a particle cannot both be accurately known at the same time. Only its most probable position or momentum can be determined.
C. Quantum mechanics includes newtonian mechanics as a special case.

8-13. Quantum Numbers

A. According to quantum theory, an electron is not restricted to a fixed orbit, but is free to move about in a three-dimensional **probability cloud**.
B. Where the probability cloud is most dense (where ψ^2 has a high value), the greatest the probability of finding the electron.
C. Three quantum numbers determine the size and shape of the probability cloud.
   1. The **principal quantum number** n governs the electron's energy and average distance from the nucleus.
   2. The **orbital quantum number** l determines the magnitude of an atomic electron's angular momentum.
   3. The **magnetic quantum number** m_l specifies the direction of an atomic electron's angular momentum.
   4. The **spin magnetic quantum number** m_s of an atomic electron has two possible values, +1/2 or -1/2, depending on whether the electron aligns itself along a magnetic field (+1/2) or opposite to the field (-1/2).

8-14. Exclusion Principle

A. The **exclusion principle**, first proposed by Wolfgang Pauli in 1925, states that only one electron in an atom can exist in a given quantum state.
B. Each atomic electron must have a different set of quantum numbers n, l, m_l, and m_s.

KEY TERMS AND CONCEPTS

The questions in this section will help you review the key terms and concepts from Chapter 8.

Multiple Choice

Circle the best answer for each of the following questions.

1. Investigation of the photoelectric effect led to the discovery that
   a. electrons exist
   b. photons have no mass and travel with the speed of light
   c. light has both wave properties and particle properties
   d. light is a form of electromagnetic energy

2. In the equation $E = hf$, h stands for
   a. light frequency
   b. Planck's constant
   c. wavelength
   d. particle momentum

3. When x-rays are produced by an x-ray tube
   a. electron KE is transformed into photon energy
   b. x-ray photon energy is transferred to electrons
   c. the effect is the same as the photoelectric effect
   d. there is a direct relationship between the number of electrons emitted in the tube and the energy of the x-rays produced

4. The momentum of a proton and an electron are calculated. The proton is shown to have more momentum than the electron; therefore, the de Broglie wavelength of the proton is
   a. longer than the de Broglie wavelength of the electron
   b. shorter than the de Broglie wavelength of the electron
   c. the same as that of the electron
   d. unable to be determined

5. Concerning an atomic electron, it is impossible to determine both its exact position and its exact ____________ at the same time.
   a. mass
   b. momentum
   c. charge
   d. quantum number

6. Which one of the following objects is not governed by the uncertainty principle?
   a. an automobile
   b. an atomic electron
   c. an atom
   d. all of these objects are governed by the uncertainty principle

7. According to the Bohr model of the atom, an atomic electron
   a. in the innermost orbit has the most energy
   b. cannot jump to an orbit of higher energy by absorbing energy
   c. can have only certain particular energies
   d. has no fixed orbit

8. The quantum number *n* of an electron orbit equals
   a. the number of electrons that can occupy the orbit
   b. the number of electron waves that fit into the orbit
   c. the speed of the electrons as they move about the nucleus
   d. the number of photons the orbital electrons emit when each jumps to a lower energy level

9. Which one of the following best describes laser-produced light?
   a. multiple-frequency incoherent light
   b. multiple-frequency coherent light
   c. single-frequency incoherent light
   d. single-frequency coherent light

10. For an electron circulating in a probability cloud, where ψ^2 has a high value
   a. the electron has the highest probability of being found
   b. the electron is least likely to be found
   c. the probability of finding the electron is zero
   d. the probability of finding the electron cannot be determined

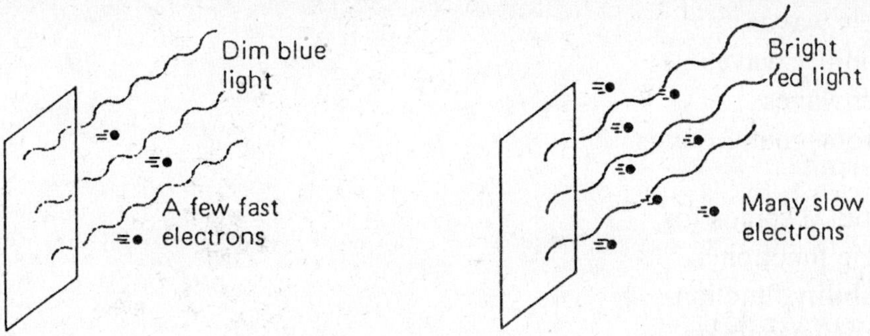

Refer to the above drawings to answer questions 11 through 13. The drawings represent light of different frequencies striking a metal surface.

11. The drawing demonstrates the
    a. electromagnetic effect
    b. photoelectric effect
    c. production of x-rays
    d. creation of electrons

12. Compared to red light, blue light
    a. travels faster than red light
    b. is dimmer than red light
    c. has a higher frequency than red light
    d. has more photons than red light

13. If the brightness of the blue light was increased
    a. fewer fast electrons would be emitted
    b. more fast electrons would be emitted
    c. there would be no change in the number of fast electrons emitted
    d. many slow electrons would be emitted

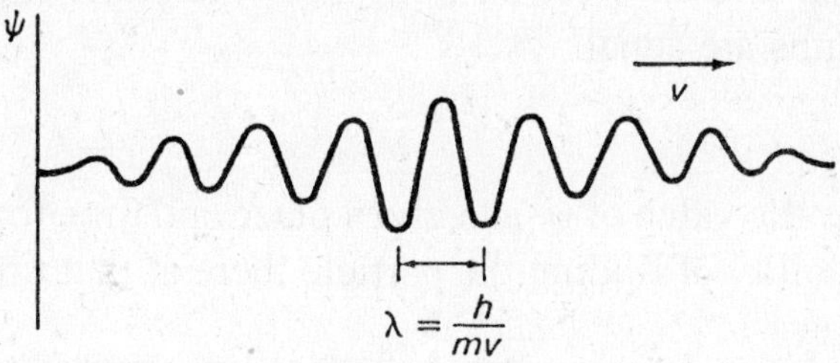

Refer to the above drawing to answer questions 14 through 16. The drawing represents the wave description of a moving object.

14. The waves associated with the wave behavior of a moving particle are called
   a. particle waves
   b. probability waves
   c. matter waves
   d. electromagnetic waves

15. The symbol ψ stands for
   a. particle function
   b. probability function
   c. de Broglie function
   d. wave function

16. The drawing indicates a wide wave packet. Which one of the following properties cannot be precisely determined?
   a. de Broglie wavelength
   b. position of the particle
   c. particle's momentum
   d. probability density

True or False

Decide whether each statement is true or false. If false, briefly state why it is false or correct the statement to make it true. See Chapters 1 or 2 for an example.

T 1. All the quanta associated with a particular frequency of light have the same energy.

T 2. Like light, electrons also exhibit particle characteristics and wave characteristics.

T 3. If a proton and an electron both have the same momentum, their de Broglie wavelengths are equal.

F 4. The larger the value of ψ^2 at a given place and time for a given particle, the lower the probability of finding the particle there at that time.

T 5. The de Broglie waves associated with a moving object travel with the same speed as the object.

F 6. The narrower the wave packet of a moving particle, the more precisely the particle's momentum can be specified.

F 7. One of the advantages ruby lasers have over helium-neon lasers is that the light of the ruby laser is produced continuously, not as separate flashes.

T 8. Newtonian mechanics is an approximate version of quantum mechanics.

F 9. The larger the principal quantum number (n) is for a given atomic electron, the closer the electron tends to be toward the nucleus.

T 10. It is impossible for any two electrons in an atom to have the same set of quantum numbers.

Fill in the Blank

1. In the photoelectric effect, electrons are emitted from a metal surface when a light beam is directed on it.

2. The quantity that varies in a matter wave is called the wave function (2 words).

3. The uncertainty principle states that it is impossible to simultaneously know both the exact position and momentum of a particle.

4. A spectroscope is a device that disperses the light emitted by an excited gas into the different frequencies the light contains.

5. An absorption spectrum is produced when white light passes from a glowing source through a cool gas.

6. In the Bohr model of the hydrogen atom, the radius of each electron orbit is proportional to the square of the orbit's quantum number.

7. An atom emits a photon when returning from an excited energy state to its ground state.

8. An atom has the lowest possible energy when it is in it ground state (2 words).

9. A laser is a device that produces an intense beam of single-frequency light.

10. The densest part of a __probability cloud__ (2 words) is where an electron is likely to be found.

11. The __principle__ quantum number governs an electron's energy and its average distance from the nucleus.

12. The __magnetic__ quantum number determines the direction of an electron's angular momentum.

13. The __orbital__ quantum number determines the magnitude of an electron's angular momentum.

14. The __spin__ magnetic quantum number governs the direction of the spin of an electron.

15. The __exclusion__ principle states that only one electron in an atom can exist in a given quantum state.

Matching

Match the name of the person on the left with the description on the right.

1.______ Max von Laue

2.______ Wolfgang Pauli

3.______ Albert Einstein

4.______ Louis de Broglie

5.______ Werner Heisenberg

6.______ Niels Bohr

8.______ Max Planck

9.______ Erwin Schrödinger

a. theory of hydrogen atoms
b. exclusion principle
c. matter waves
d. quantum theory of light
e. quanta
f. identified x-rays as high frequency em waves
g. his equation is mathematical basis of quantum mechanics
h. uncertainty principle

SOLVED PROBLEMS

Study the following solved example problem as it will provide insight into solving the problems listed at the end of Chapter 8 in *The Physical Universe.* Review the mathematics refresher in the Appendix of the Study Guide if you are unfamiliar with the basic mathematical operations presented in this example.

Example 8-1

A laser emits a beam of light having a wavelength of 650 nm. What is the quantum energy, in joules, of the photons?

Solution

Remember that all the quanta (photons) associated with a particular frequency of light have the same energy. The equation to use is

$$E = hf$$

Quantum energy = (Planck's constant)(frequency)

To solve the problem, the value for f must be determined. We can use the formula for wavelength (from Table 6.1) to determine the value for f

$$\lambda = v/f$$

where λ = wavelength, v = speed, and f = frequency in Hz.

Solving for f, the equation becomes

$$f = v/\lambda$$

The problem states the wavelength is 650 nm which is 6.50×10^{-7} m. The value of v is the speed of light, or 3.00×10^{8} m/s. Substituting these values into the equation for frequency, we get

$$f = v/\lambda = (3.00 \times 10^{8}\text{ m/s})/(6.50 \times 10^{-7}\text{ m}) = 4.62 \times 10^{14}\text{ Hz}$$

We can determine the quantum energy of the photons of the laser beam by substituting the values for Planck's constant and frequency into the equation for quantum energy:

$$E = hf$$

$$E = (6.63 \times 10^{-34}\text{ J} \bullet \text{s})(4.62 \times 10^{14}\text{ Hz})$$

$$E = 1.37 \times 10^{-19}\text{ J}$$

WEB LINKS

Investigate the photoelectric effect at

http://webphysics.ph.msstate.edu/jc/library/27-4/simulation.html

Investigate Bohr's model of the atom at

http://www.walter-fendt.de/ph11e/bohrh.htm

ANSWER KEY

Multiple Choice

1. c 2. b 3. a 4. b 5. b 6. d 7. c 8. b 9. d 10. a 11. b 12. c 13. b 14. c 15. d 16. b

True or False

1. True
2. True
3. True
4. False. The larger the value of ψ^2 at a given place and time for a given particle, the <u>higher</u> the probability of finding the particle there at that time.
5. True
6. False. The narrower the wave packet of a moving particle, the more precisely the particle's <u>position</u> can be specified.
7. False. Helium-neon lasers operate continuously; ruby lasers produce separate flashes.
8. True
9. The larger the principal quantum number (n) is for a given atomic electron, the farther the electron tends to be from the nucleus.
10. True

Fill in the Blank

1. photoelectric
2. wave function
3. uncertainty
4. spectroscope
5. absorption
6. quantum
7. photon
8. ground state
9. laser
10. probability cloud
11. principal
12. magnetic
13. orbital
14. spin
15. exclusion

Matching

1. g 2. c 3. e 4. d 5. i 6. b 7. a 8. f 9. h

Chapter 9

THE PERIODIC LAW

OUTLINE

GOALS

1. Distinguish among the three classes of matter: elements, compounds, and mixtures.
2. State the law of definite proportions.
3. Explain the meanings of the letters, numbers, and parentheses in the chemical formula of a compound, for instance, $Al_2(SO_4)_3$.
4. Compare the properties of metals and nonmetals.
5. List some of the characteristic properties of the halogens, the alkali metals, and the inert gases.
6. State the periodic law and describe how the periodic table is drawn up.
7. Distinguish between the groups and periods of the periodic table.
8. State what is meant by atomic shells and subshells.
9. Distinguish between metal and nonmetal atoms in terms of their electron structures.
10. Explain the origin of the periodic law in terms of the electron structures of atoms.
11. Compare covalent and ionic bonds.
12. State what is meant by a polar covalent molecule.
13. Discuss the nature of an atom group.
14. Explain what a chemical equation represents and does not represent.
15. Recognize whether a chemical equation is balanced or unbalanced.
16. Balance an unbalanced chemical equation.

CHAPTER SUMMARY

Chapter 9 introduces the science of chemistry with a discussion of what is meant by **chemical reaction** and chemical change. The terms **element, compound**, and **mixture** are defined, and the **law of definite proportions** is presented. The information contained in a **chemical formula** is discussed. The physical properties of metals and nonmetals are given, and the basis for their chemical behavior is presented. The **periodic table** is introduced, and the quantum theory of the atom of Chapter 8 is shown to be the basis of the **periodic law**. The chemical behavior of atoms is related to their electron structure, and the concepts of atomic shells and subshells are presented. **Covalent bonds** and **ionic bonds** are examined, and the rules for naming compounds are given. The information contained in a **chemical equation** is discussed.

CHAPTER OUTLINE

9-1. Chemical Change

A. The medieval search for a way to change ordinary metals into gold is called **alchemy**.
B. A **chemical reaction** results in the formation of a new substance whose properties are different from those of the individual substances that participate in the reaction.
C. A **heterogeneous substance** is composed of separate materials of different kinds that are not chemically combined and retain their individual properties.

9-2. Three Classes of Matter

A. Three classes of matter are **elements**, **compounds,** and **mixtures**.
   1. An element cannot be broken down or changed into another element by chemical means. All other substances are combinations of elements.
   2. Compounds are formed when two or more elements combine chemically to produce a new substance. The properties of a compound are different from those of the elements it contains.
   3. Mixtures consist of elements or compounds (or both) that are not chemically combined and retain their characteristic properties.
      a. **Heterogenous mixtures** are those whose constituents are not uniformly mixed and are easy to separate.
      b. **Homogeneous mixtures**, or **solutions**, are those whose constituents are uniformly mixed.

B. Compounds can be distinguished from solutions by two tests:
   1. Compounds boil and freeze at definite temperatures and are not altered by a change of state. Mixtures lack definite boiling and freezing temperatures and can be altered by a change of state.
   2. In a compound, the elements are present in a specific ratio by mass according to the **law of definite proportions**. In a mixture, the components are not present in a specific ratio by mass.

9-3. The Atomic Theory

A. The English schoolteacher John Dalton (1766-1844) proposed an atomic theory which provided the basis for our modern understanding of matter.
   1. Dalton stated that all matter was composed of basic particles called **atoms**.
   2. Dalton was the first to establish the relative masses of the atoms of different elements.

B. According to the modern atomic theory of matter:
   1. Atoms are the ultimate particles of any element.
   2. **Molecules** are the ultimate particles of a gaseous compound.
   3. The ultimate particles of liquid and solid compounds may be atoms, molecules, or ions.

C. A **chemical formula** indicates the number of atoms and types of elements present in a compound.
   1. The kinds of atoms in a compound are represented by their chemical symbols.
   2. The numbers of each kind of atom are represented by subscript numbers.

9-4. Metals and Nonmetals

A. Most elements are metals.

B. All metals, except mercury, are solid at room temperature. Nonmetals may be solid, liquid, or gaseous.

C. Metals have a characteristic **metallic luster**; nonmetals do not.

D. All metals are opaque; most nonmetals are transparent in thin sheets.

E. Metals can be shaped by bending or hammering. Solid nonmetals are brittle.

F. Metals are good conductors of heat and electricity. Nonmetals are insulators.

G. Some elements, like carbon, have properties that are transitional between metals and nonmetals.

9-5. Chemical Activity

A. **Active** elements combine readily to form compounds. **Inactive** elements have little tendency to react chemically.

B. Active elements liberate more heat when they react than do inactive elements.

C. Active elements usually form stable compounds.

9-6. Families of Elements

A. Elements which resemble one another in the way they react chemically are said to belong to the same family of elements.

B. Some examples of chemical families include:
   1. The **halogens**, or "salt formers," are active nonmetals.
   2. The **alkali metals** are active metals and have low melting points.

3. The **inert gases** are inactive nonmetals.

9-7. The Periodic Table

A. The Russian chemist Dmitri Mendeleev formulated the **periodic law** about 1869 which states that when elements are listed in order of atomic number, elements with similar chemical and physical properties appear at regular intervals.

B. The **periodic table** is a listing of the elements according to atomic number in a series of rows such that elements with similar properties form vertical columns.

9-8. Groups and Periods

A. The periodic table arranges chemical families of elements in vertical columns called **groups**.

B. The horizontal rows of elements of the periodic table are called **periods**.

C. The **transition elements** are placed between groups 2 and 3 and include:
   1. The **rare-earth** metals (atomic numbers 57-71).
   2. The **actinides** (atomic numbers 89-105).

D. Mendeleev's periodic table, although arranged by increasing atomic mass rather than atomic number, helped him predict the existence and properties of a number of elements.

9-9. Shells and Subshells

A. The electrons in an atom that have the same principal quantum number n occupy the same **shell**.

B. The electrons in an atom that have the same orbital quantum number l occupy the same **subshell**. The larger the value of l, the more electrons the subshell can hold.

C. A shell or subshell that contains its full quota of electrons is said to be closed.

9-10. Explaining the Periodic Table

A. The electrons in a closed shell or subshell are tightly bound to the atom and not easily removed.

B. The inert gases all have closed subshells and are thus chemically inactive.

C. Metal atoms have one or several electrons outside closed shells or subshells and combine chemically by losing these electrons to nonmetal atoms.

D. Nonmetal atoms need one or several electrons to achieve closed shells or subshells and combine chemically by taking electrons from metal atoms or by sharing electrons with nonmetal atoms.

E. Across any period of the periodic table, metallic activity (losing electrons) decreases from left to right and nonmetallic activity (gaining electrons) increases from left to right.

9-11. Types of Bond

A. A **covalent bond** is formed when one or more pairs of electrons are shared by two or more atoms.

B. An **ionic bond** is formed when electrons are transferred between two or more atoms and the resulting ions of opposite charge attract each other.

C. Covalent bonds form molecules. Ionic bonds usually do not form molecules but aggregates of ions attracted by charge.

9-12. Covalent Bonding

A. **Covalent compounds** are substances whose atoms are joined by one or more pairs of electrons in a covalent bond.

B. **Polar covalent compounds** are those in which the shared electron pairs are closer to one atom than to the other, making one part of the molecule relatively negative and another part relatively positive.

C. **Organic compounds** are covalent compounds that contain the element carbon.

9-13. Ionic Bonding

A. **Ionic compounds** are formed by electron transfer and are usually crystalline solids with high melting points.

B. Some ionic compounds contain a metal and a single nonmetal; others contain a metal and an **atom group** consisting of two or more nonmetals.

9-14. Ionic Compounds

A. When a metal atom combines with a nonmetal atom to form an ionic compound, the chemical formula of the ionic compound formed can be determined by knowing how many electrons the metal atom loses and how many electrons the nonmetal atom gains.

B. To write the formula of an ionic compound consisting of a metal and a nonmetal:

1. Determine the ionic charge of the metal and of the nonmetal.
2. Write the value of the ionic charge of the nonmetal as a subscript following the metal.
3. Write the value of the ionic charge of the metal as a subscript following the nonmetal.
4. If the value of the ionic charge for either the metal or the nonmetal is 1, no subscript is written.

9-15. Atom Groups

A. Atom groups appear as units in many compounds and remain together during chemical reactions.
B. The **sulfate group** SO_4 is an example of an atom group.
C. A **precipitate** is an insoluble solid that results from a chemical reaction in solution.
D. When two or more atom groups of the same kind are present in the formula of a compound, parentheses are placed around the group.

Example: $Ca(NO_3)_2$

9-16. Naming Compounds

A. A compound ending in -ide usually is composed of only two elements. Hydroxides which contain the OH^- ion are an exception.
B. A compound ending in -ate contains oxygen and two or more other elements.
C. When the same pair of elements occurs in two or more compounds, a prefix (*mono* = 1, *di* = 2, *tri* = 3, *tetra* = 4, *penta* = 5, *hexa* =6, and so on) may be used to indicate the number of one or both kinds of atoms in the molecule.
D. When one of the elements in a compound is a metal that can form different ions, the ionic charge of the metal is given by a roman numeral.

9-17. Chemical Equations

A. A **chemical equation** expresses the results of a chemical change.
B. In a chemical equation the formulas of the reacting substances (**reactants**) appear on the left-hand side and the formulas of the products appear on the right-hand side.
C. Chemical equations must be **balanced**, meaning that the number of atoms of each kind of element must be the same on both sides of the equation. **Unbalanced** chemical equations have unequal numbers of at least one kind of atom on both sides of the equation.
D. Unbalanced chemical equations are balanced by placing numbers, or coefficients, in front of the reactants and products to adjust their amounts.
    1. The numerical coefficient multiplies the number of each kind of atom in the formula.
    2. The subscripts of the formulas of the reactants and products are never changed when balancing a chemical equation since this changes the identity of the substances.

KEY TERMS AND CONCEPTS

The questions in this section will help you review the key terms and concepts from Chapter 9.

Multiple Choice

Circle the best answer for each of the following questions.

1. A solution is a
   a. heterogeneous substance
   b. chemical compound
   c. homogeneous mixture
   d. pure substance

2. Which one of the following is not characteristic of a compound?
   a. has a specific boiling point and freezing point
   b. is not chemically altered by a change in state
   c. its elements are present in a specific ratio
   d. cannot be broken down into two or more elements by chemical means

3. The person who first established the relative masses of different elements was
   a. Robert Boyle
   b. Jean Rey
   c. Georg Stahl
   d. John Dalton

4. The name halogen means
   a. inert gas
   b. salt former
   c. bad odor
   d. acid former

5. The alkali metals
   a. have relatively high melting points
   b. form unstable compounds with nonmetals
   c. are inactive chemically
   d. react with dilute acids to liberate hydrogen

6. From left to right, each period of the periodic table terminates in a(n)
   a. inert gas
   b. active metal
   c. weakly active nonmetal
   d. highly active nonmetal

7. The rare-earth and actinide elements belong to the
   a. halogens
   b. transition metals
   c. inert gases
   d. alkali metals

8. Across any period of the periodic table
   a. metallic activity decreases to the right
   b. metallic activity increases to the right
   c. metallic activity remains constant to the right
   d. nonmetallic activity decreases to the right

9. Organic compounds are covalent compounds that all contain the element
   a. oxygen
   b. nitrogen
   c. carbon
   d. sulfur

10. The hydroxides contain the
   a. NH_4^+ ion
   b. NO_3^- ion
   c. SO_4^{2-} ion
   d. OH^- ion

11. In the chemical formula KNO_3, the subscript 3 represents
   a. 3 molecules of KNO_3
   b. 3 oxygen atoms
   c. 3 NO atom groups
   d. 1/3 atom of oxygen

True or False

Decide whether each statement is true or false. If false, either briefly state why it is false or correct the statement to make it true. See Chapter 1 or 2 for an example.

T 1. The molecules of a compound have fixed compositions according to the law of definite proportions.

T 2. In general, the more active an element is, the more stable are its compounds.

F 3. The elements in a period have similar properties.

T 4. A metal atom has one or several electrons outside closed shells or subshells and combines chemically by losing these electrons to nonmetal atoms.

T 5. In a polar covalent compound, the atoms share electrons to an unequal extent.

T 6. Most ionic compounds are crystalline solids with high melting points.

F 7. The correct chemical name for NaBr is sodium bromate.

F 8. The correct chemical name for $Ca(OH)_2$ is calcium(II) hydroxide.

T 9. The correct chemical name for Al_2O_3 is aluminum oxide.

T 10. The correct chemical name for $HgCl_2$ is mercury(II) chloride.

Fill in the Blank

1. The rusting of iron and the burning of wood are examples of chemical reactions.
2. Two or more elements may chemically combine to form a compound.
3. A mixture has a variable composition whose parts can be separated by physical means.
4. A uniform mixture of different substances is known as a solution.
5. According to the law of definite proportions, elements combine in a specific mass ratio when they form a compound.
6. Molecules are the ultimate particles of a gaseous compound.
7. A periodic table is a listing of the elements according to atomic number.

8. Atomic electrons with the same quantum number n occupy the same ______________.

9. A ______________ bond is formed by the sharing of electron pairs.

10. In order to correctly represent a chemical reaction, a chemical equation must be ______________.

Matching

Match the term on the left with its definition on the right.

1.____	metallic luster	a. "salt formers"
		b. the sheen of a clean metal surface
2.____	halogens	c. horizontal rows of elements of the periodic table
3.____	periods	d. vertical columns of the elements of the periodic table
4.____	ionic compounds	e. chemically inactive elements
		f. compounds formed by electron transfer
5.____	precipitate	g. compounds formed by the sharing of electron pairs
6.____	inert gases	h. insoluble solid resulting from a chemical reaction in solution
7.____	transition elements	i. include the rare-earth metals and the actinides
8.____	covalent compounds	j. cannot be broken down into new substances by chemical means
9.____	element	
10.____	groups	

SOLVED PROBLEMS

Study the following solved example problems as they will provide insight into solving the problems listed at the end of Chapter 9 in *The Physical Universe*. Review the mathematics refresher in the Appendix of the Study Guide if you are unfamiliar with the basic mathematical operations presented in these examples.

Example 9-1

With the help of Tables 9-8 and 9-9 in *The Physical Universe*, find the formulas of the following compounds: potassium cyanide; calcium chloride; sodium carbonate; aluminum phosphate.

Solution

All of the compounds are ionic and are easily named by using the criss-cross method. The following steps outline this method using the compounds given in the problem as examples.

1. Write the symbol of the positive ion first, and then the negative ion.

potassium cyanide	K^+	CN^-
calcium chloride	Ca^{2+}	Cl^-
sodium carbonate	Na^+	CO_3^{2-}
aluminum phosphate	Al^{3+}	PO_4^{3-}

2. Write a subscript number equal to the magnitude of the ionic charge of the other ion. An ion having either a + or − sign is understood to have a subscript of 1. Enclose atom group ions in parentheses for now.

potassium cyanide	K_1^+	$(CN^-)_1$
calcium chloride	Ca_1^{2+}	Cl_2^-
sodium carbonate	Na_2^+	$(CO_3^{2-})_1$
aluminum phosphate	Al_3^{3+}	$(PO_4^{3-})_3$

3. Rewrite the formulas omitting the ionic charges, the subscript 1, and the parentheses of the atom group ions which have the subscript 1.

potassium cyanide	KCN
calcium chloride	$CaCl_2$
sodium carbonate	Na_2CO_3
aluminum phosphate	$Al_3(PO_4)_3$

4. Reduce the subscript numbers in the final formula to their lowest terms.

potassium cyanide	KCN
calcium chloride	$CaCl_2$
sodium carbonate	Na_2CO_3
aluminum phosphate	$AlPO_4$

Example 9-2

Name these compounds: AgCl; $Zn(MNO_4)_2$; Al_2O_3; $Cu(OH)_2$; $NaNO_3$

Solution

AgCl is named silver chloride because the compound contains only two elements.

$Zn(MNO_4)_2$ is named zinc permanganate because compounds that contain oxygen and two or more other elements end in -ate. Remember that MNO_4^- is the permanganate atom group.

Al_2O_3 is named aluminum oxide because the compound contains only two elements.

$Cu(OH)_2$ is named copper(II) hydroxide. Copper forms two possible ions (Cu^+, Cu^{2+}). The subscript 2 after OH in the formula states there are two OH^- ions, thus the copper ion in the compound must be Cu^{2+} in order for $Cu(OH)_2$ to be electrically balanced. The Roman numeral II is written in parentheses after Cu in the formula to indicate the Cu^{2+} ion; therefore, $Cu(OH)_2$ is named copper(II) hydroxide because of the presence of the Cu^{2+} ion and the hydroxide group.

$NaNO_3$ is named sodium nitrate because NO_3 is the nitrate group.

Example 9-3

Write a balanced equation for the reaction that occurs when silver nitrate ($AgNO_3$) reacts with potassium chloride (KCl) to form silver chloride (AgCl) and potassium nitrate (KNO_3).

Solution

1. Write the reaction in the form of a chemical equation using chemical formulas.

   Silver nitrate ($AgNO_3$) and potassium chloride (KCl) are the reactants and are written on the left side of the equation. Silver chloride (AgCl) and potassium nitrate (KNO_3) are the products and are written on the right side of the equation.

   $$AgNO_3 + KCl \rightarrow AgCl + KNO_3$$

2. Determine the number of atoms of each element on both sides of the equation.

   On the reactant side there are: one Ag atom, one N atom, three O atoms, one K atom, and one Cl atom. On the product side there are: one Ag atom, one N atom, three O atoms, one K atom, and one Cl atom

3. Balance the equation, if necessary.

Since there are equal numbers of each kind of atom on both sides of the equation, the equation is balanced.

Example 9-4

Burning magnesium in an atmosphere of pure carbon dioxide produces magnesium oxide (MgO) and carbon (C). Write a balanced equation for this reaction.

Solution

1. Write the reaction in the form of a chemical equation using chemical formulas.

   Magnesium (Mg) and carbon dioxide (CO_2) are the reactants. Magnesium oxide (MgO) and carbon (C) are the products. The chemical equation is:

$$Mg + CO_2 \rightarrow MgO + C$$

2. Determine the number of atoms of each element on both sides of the equation.

   There are one Mg atom and one C atom on each side of the equation; however, there are two O atoms on the reactant side and one O atom on the product side. The equation is unbalanced.

3. Balance the equation, if necessary.

   Since the number of O atoms is not the same on both sides of the equation, a logical first step would be to place a numerical coefficient of 2 in front of the MgO on the product side to balance the number of O atoms on each side of the equation. The equation now reads:

$$Mg + CO_2 \rightarrow 2MgO + C$$

   Notice that there is one Mg atom on the reactant side, and there are two Mg atoms on the product side. The equation is balanced by placing a coefficient of 2 in front of the Mg on the reactant side. The balanced equation is:

$$2Mg + CO_2 \rightarrow 2MgO + C$$

WEB LINKS

Virtual periodic tables can be seen at

http://www.thinkquest.org/library/site_sum.html?lib_id=504&team_id=12909

http://www.webelements.com/index.html

http://chemlab.pc.maricopa.edu/periodic/periodic.html

Practice balancing chemical equations at these sites. Both sites require the Macromedia Shockwave Player.

http://www.wfu.edu/~ylwong/balanceeq/balanceq.html

http://www.mpcfaculty.net/mark_bishop/balancing_equations_tutorial.htm

Chemtutor is a great resource for help with the chemistry material presented in this chapter, and chapters 10, 11, and 12

http://chemtutor.com

Another useful site is General Chemistry Online at

http://antoine.frostburg.edu/chem/senese/101.index.shtml

ANSWER KEY

Multiple Choice

1. c 2. d 3. d 4. b 5. d 6. a 7. b 8. a 9. c 10. d 11. b

True or False

1. True
2. True
3. False. The elements in a <u>group</u> have similar properties.
4. True
5. True
6. True
7. False. The correct chemical name for NaBr is sodium bromide since the ending -ide usually indicates a compound with only two elements.
8. False. The correct chemical name for $Ca(OH)_2$ is calcium hydroxide since calcium metal forms only a single ion, Ca^{2+}.
9. True
10. True

Fill in the Blank

1. chemical
2. compound
3. mixture
4. solution
5. compound
6. gaseous
7. periodic
8. shell
9. covalent
10. balanced

Matching

1. b 2. a 3. c 4. f 5. h 6. e 7. i 8. g 9. j 10. d

Chapter 10
CRYSTALS, IONS, AND SOLUTIONS

OUTLINE

GOALS

1. Distinguish between crystalline and amorphous solids.
2. List the four classes of crystalline solids and identify the nature of the bonds in each class.
3. Explain the origin of van der Waals forces.
4. Distinguish between solvent and solute.
5. Define the solubility of a substance.
6. Describe what is meant by unsaturated, saturated, and supersaturated solutions.
7. Discuss how the solubilities of gases and solids in water vary with temperature and, in the case of gases, with pressure.
8. Compare the behavior of solvents of polar and nonpolar liquids.
9. Give some of the reasons why an ionic crystal is believed to dissociate into ions when it dissolves.
10. Explain how dissociation occurs.
11. Discuss some of the chief causes of water pollution.
12. Define acid and distinguish between strong and weak acids.
13. Define base and distinguish between strong and weak bases.
14. Describe the pH scale.
15. Explain what happens when an acid and a base neutralize each other.
16. Describe how to prepare a salt.
17. Give some examples of acids, bases, and salts.

CHAPTER SUMMARY

Chapter 10 discusses crystals, ions, and solutions. The physical properties of **ionic**, **covalent**, **metallic,** and **molecular** crystalline solids are presented, and the chemical bond responsible for each crystal type is discussed. The concept of **solubility** is introduced, and the terms **solvent** and **solute** are defined. The physical basis for the behavior of **polar liquids** and **nonpolar liquids** is examined, and the **dissociation** of a compound into ions and the formation of **electrolytes** is discussed. The properties of ions in solution are explored, leading to a discussion of the properties of **acids** and **bases**. The **pH scale** is introduced, and **neutralization reactions** and the formation of **salts** are discussed.

CHAPTER OUTLINE

10-1. Ionic and Covalent Crystals

A. Most solids are **crystalline**, meaning the particles that compose them are arranged in repeated patterns.
B. **Amorphous** solids have particles irregularly arranged.
C. Crystalline solids fall into four classes:
   1. Ionic
   2. Covalent
   3. Metallic
   4. Molecular
D. Ionic crystals are formed by the attraction between positive and negative ions.
E. Covalent crystals are formed when pairs of electrons are shared between adjacent atoms.
F. Some crystals are neither wholly ionic nor wholly covalent but contain bonds of mixed character.
G. Diamond and **graphite** are forms of carbon but differ from each other in their physical properties.

10-2. The Metallic Bond

A. The **metallic bond** is formed by a "gas" of electrons that moves freely through the assembly of metal ions that form a solid metal.
B. The metallic bond accounts for the characteristics of metals.

10-3. Molecular Crystals

A. Some liquids and solids are formed through the action of **van der Waals forces**, named after the Dutch physicist Johannes van der Waals.
B. There are several types of van der Waals interactions:
   1. Polar-polar interaction occurs between **polar molecules** whose positively and negatively charged ends cause them to line up with the ends that have opposite

charges adjacent.

2. Polar-nonpolar interaction occurs between polar and **nonpolar molecules** because the electric field of the polar molecules causes separations of charge in the nonpolar molecules. The oppositely charged ends of the polar and nonpolar molecules produce an attractive force.
3. Nonpolar-nonpolar interaction occurs between nonpolar molecules when the molecule's electrons at any given moment are distributed unevenly. This creates temporarily charged molecules whose adjacent ends having opposite signs results in an attractive force.

C. Van der Waals forces are much weaker than ionic, covalent, and metallic bonds.

10-4. Solubility

A. In a solution, the substance present in larger amount is the **solvent**; the other is the **solute**.

B. The **concentration** of a solution is the amount of solute in a given amount of solvent.

C. The **solubility** of a substance is the maximum amount that can be dissolved in a given quantity of a particular solvent at a given temperature.

D. A **saturated** solution contains the maximum amount of solute possible at a given temperature; a **supersaturated** solution contains more dissolved solute than is normally possible at a given temperature and is usually unstable.

E. The solubilities of solids increase with increasing temperatures, while the solubilities of gases in liquids decrease with increasing temperatures.

F. The boiling point of a solution is usually higher than that of the pure solvent, and its freezing point is lower.

10-5. Polar and Nonpolar Liquids

A. A **polar liquid** is a substance whose molecules behave as if negatively charged at one end and positively charged at the other. The molecules of a **nonpolar liquid** have uniform charge distributions.

B. Polar liquids dissolve only ionic and polar covalent compounds. Nonpolar liquids dissolve only nonpolar covalent compounds.

C. **Dissociation** refers to the separation of a compound into ions when it dissolves.

D. **Electrolytes** are substances that dissociate into ions when dissolved in water; **nonelectrolytes** are soluble covalent compounds that do not dissociate in solution.

E. Electrolytes in solution are able to conduct electric current.

10-6. Ions in Solution

A. Ions in solution have their own sets of properties that differ from their original atoms and from the original solute. Dissociation is a type of chemical change.

B. The properties of a solution of an electrolyte are the sum of the properties of the ions present in the solution.

10-7. Evidence for Dissociation

A. In 1887, the Swedish chemist Svante Arrhenius proposed that many substances exist as ions in solution.
B. Arrhenius based his hypothesis on two points:
   1. Reactions between electrolytes in solution occur almost instantaneously, but very slowly or not at all if the electrolytes are dry.
   2. Electrolyte solutions have lower freezing points than comparable solutions of nonelectrolytes.

10-8. Water

A. Seawater has an average salt content, or **salinity**, of 3.5 percent.
B. "Hard" water is freshwater that contains Ca^{2+} and Mg^{2+} ions in solution; "soft" water is free of Ca^{2+} and Mg^{2+} ions.

10-9. Water Pollution

A. Sources of water pollution include:
   1. Industrial pollutants
   2. Agricultural fertilizers and pesticides
   3. Thermal pollution
B. The **biochemical oxygen demand**, or BOD, is the amount of oxygen needed to completely oxidize the organic material in a sample of water.

10-10. Acids

A. An **acid** is a substance that contains hydrogen and whose solution in water increases the number of H^+ ions present.
B. The H^+ ions released when an acid dissociates in water combine with water molecules to produce **hydronium ions,** H_3O^+.
C. The characteristic properties of acids are actually the properties of H_3O^+ ions rather than the properties of H^+ ions.
D. The water solutions of acids taste sour, and acids change the color of litmus dye from blue to red.

10-11. Strong and Weak Acids

A. **Strong acids** dissociate completely; **weak acids** dissociate only slightly.

B. Some substances, such as carbon dioxide, do not contain hydrogen but produce acidic solutions by reacting with water to liberate H^+ ions from water molecules.

10-12. Bases

A. A **base** is a substance that contains hydroxide groups and whose solution in water increases the number of OH^- ions present.
B. **Strong bases** dissociate completely; **weak bases** dissociate only slightly.
C. Some substances, such as ammonia, do not contain OH but produce basic solutions because they react with water to release OH^- ions from water molecules.
D. The water solutions of bases have a bitter taste, a soapy feel, and turn red litmus to blue.
E. The name **alkali** is sometimes used for a substance that dissolves in water to give a basic solution. The terms alkaline and basic mean the same.

10-13. The pH Scale

A. Pure water dissociates very slightly into H^+ and OH^- ions.
   1. In an acidic solution, the concentration of H^+ ions is greater than in pure water, and the concentration of OH^- ions is lower.
   2. In a basic solution, the concentration of OH^- ions is greater than in pure water, and the concentration of H^+ ions is lower.
B. The **pH scale** expresses the exact degree of acidity or basicity of a solution in terms of its H^+ ion concentration.
   1. A solution that is neither acidic nor basic is said to be **neutral** and has a pH of 7.
   2. Acidic solutions have pH values of less than 7.
   3. Basic solutions have pH values of more than 7.
C. A change in pH of 1 means a change in H^+ ion concentration by a factor of 10.

10-14. Salts

A. When a basic solution is mixed with an acidic solution, the base destroys, or **neutralizes**, the properties of the acid and vice versa. The process is called **neutralization**.
B. In neutralization reactions, H^+ and OH^- ions join to form water molecules.
C. Ions left in solution as a result of neutralization can combine to form a **salt** when the solution is evaporated to dryness.
D. Most salts are crystalline solids that consist of positive metal ions and negative nonmetal ions.

KEY TERMS AND CONCEPTS

The questions in this section will help you review the key terms and concepts from Chapter 10.

Multiple Choice

Circle the best answer for each of the following questions.

1. Crystalline solids are those whose particles
   a. are extremely hard
   b. are arranged in repeated patterns
   c. are shaped like tiny crystals
   d. are without form

2. In covalent crystals, the atoms are held together by
   a. shared electron pairs between adjacent atoms
   b. the stable assembly of positive and negative ions
   c. electron transfer between metal and nonmetal atoms
   d. an electron "gas" that moves freely within the crystal

3. Of the following types of chemical bonds, the weakest is
   a. ionic
   b. covalent
   c. metallic
   d. van der Waals

4. Snowflakes and ice are examples of _______________ crystals.
   a. covalent
   b. ionic
   c. metallic
   d. molecular

5. The solubility of sodium chloride, or NaCl, in water is 36 g per 100 g of water at 20°C. NaCl is called the
   a. solvent
   b. concentrate
   c. solute
   d. precipitate

6. House hold water softeners use ion-exchange resins to remove Ca^{2+} and Mg^{2+} ions and replace them with
   a. Na^+ ions
   b. Ba^{2+} ions
   c. K^+ ions
   d. Al^{3+} ions

7. A water solution of a certain substance has a sour taste and turns blue litmus to red. The substance is most likely to be
   a. a base
   b. an alkali
   c. an acid
   d. a salt

8. Acids react with metals to liberate
   a. hydrogen ions
   b. hydrogen gas
   c. hydronium ions
   d. hydroxide ions

9. Of the following acids, the weakest is
   a. hydrochloric
   b. carbonic
   c. sulfuric
   d. nitric

10. Bases consist of a metal together with one or more
    a. ammonium groups
    b. hydronium ions
    c. hydrogen ions
    d. hydroxide groups

11. An acidic solution having a pH of 2 is how many times more acidic than an acidic solution of pH 6?
    a. 40
    b. 100
    c. 1,000
    d. 10,000

12. A strong base can be neutralized by combining it with a(n)
   a. strong acid
   b. salt
   c. alkali metal
   d. weak base

True or False

Decide whether each statement is true or false. If false, either briefly state why it is false or correct the statement to make it true. See Chapter 1 or 2 for an example.

___T___ 1. A crystalline solid, unlike an amorphous solid, melts at a specific temperature.

___T___ 2. Metals are good conductors of heat and electricity because their outer shell electrons can move freely through the assemblies of metal ions.

___F___ 3. Chemical bonds formed by van der Waals forces are much stronger than ionic, covalent, and metallic bonds.

___F___ 4. Molecular crystals are characterized as being very hard and having high melting points.

___F___ 5. As a general rule, the solubilities of most solids decrease with increasing temperature.

___T___ 6. Sugar is considered to be a nonelectrolyte because it does not dissociate into ions when dissolved in water.

___T___ 7. When an electrolyte such as NaCl is dissolved in water, the properties of the resulting solution are really the sum of the properties of the resulting Na^+ and Cl^- ions.

___T___ 8. The salinity of seawater averages 3.5 percent.

F 9. Pure acids are composed of ions and dissociate into their component ions when dissolved in polar liquids like water.

F 10. The biochemical oxygen demand of an aquatic system represents the amount of oxygen in solution needed by aquatic organisms to support life.

Fill in the Blank

1. The ~~molecular~~ metallic bond is formed by a "gas" of electrons that moves freely through the assembly of metal ions in a solid metal.

2. A polar liquid has molecules that behave as if positively charged at one end and negatively charged at the other.

3. Nonpolar compounds dissolve only in nonpolar liquids.

4. A saturated solution contains the maximum amount of solute possible.

5. An electrolyte such as NaCl in solution conducts electric current through the motions of its ions.

6. The average salt content, or salinity, of seawater is 3.5 percent.

7. The BOD (3 words) is the amount of oxygen needed to completely oxidize the organic material in a sample of water.

8. Strong acids or bases dissociate completely.

9. A neutral solution has a pH of 7.

10. The color of litmus paper is red in an acid solution and blue in a basic solution.

Matching

Match the terms on the left with their definitions on the right.

1._____ pH scale

2._____ electrolyte

3._____ acid

4._____ base

5._____ solvent

6._____ dissociation

7._____ salt

8._____ amorphous

9._____ crystalline

10._____ hydronium ions

a. increases the number of OH^- ions when dissolved in water

b. the separation of a compound into ions when it dissolves

c. forms when H^+ ions combine with water molecules in a solution

d. "without form"

e. substances that separate into free ions in water

f. product of a neutralization reaction

g. increases the number of H^+ ions when dissolved in water

h. measures degree of acidity or alkalinity

i. substance present in largest amount in a solution

j. having particles arranged in repeated patterns

SOLVED PROBLEMS

Study the following solved example problems as they will provide insight into solving the problems listed at the end of Chapter 10 in *The Physical Universe*. Review the mathematics refresher in the appendix of the text if you are unfamiliar with the basic mathematical operations presented in these examples.

Example 10-1

A popular antacid contains the strong base aluminum hydroxide. Give the ionic equation for the neutralization of aluminum hydroxide by stomach acid (hydrochloric acid).

Solution

First, write the equation using the chemical formulas for the known reactants and products. Since a neutralization reaction between a basic solution and an acidic solution produces a salt (in solution) and water, the equation is

$$Al(OH)_3 + HCl \rightarrow \text{salt} + H_2O$$

Second, determine the identity of the salt. $Al(OH)_3$ is a strong base and HCl is a strong acid; therefore, both $Al(OH)_3$ and HCl dissociate completely to produce the following ions:

$$Al^{3+} + OH^- + H^+ + Cl^- \rightarrow \text{salt} + H_2O$$

The H^+ and OH^- ions combine to form the water on the product side. The remaining Al^+ and Cl^- ions combine to form the salt aluminum chloride, or $AlCl_3$. The equation now reads:

$$Al(OH)_3 + HCl \rightarrow AlCl_3 + H_2O$$

Third, examine the equation to determine if it is balanced. Although the number of Al atoms on both sides of the equation is the same, the number of O, H, and Cl atoms is not. The equation is balanced by placing a coefficient of 3 in front of the HCl on the reactant side and a coefficient of 3 in front of the H_2O on the reactant side. The balanced equation is:

$$Al(OH)_3 + 3HCl \rightarrow AlCl_3 + 3H_2O$$

Example 10-2

What salt is formed when a solution of sodium hydroxide is neutralized by sulfuric acid? Give the equation of the process.

$2NaOH + H_2SO_4 \rightarrow 2H_2O + Na_2SO_4$

Solution

As in Example 10-1, this is a neutralization reaction in which a strong base and a strong acid react to form a salt and water. Substituting the formulas for the know reactants and products, the equation is

$$NaOH + H_2SO_4 \rightarrow \text{salt} + H_2O$$

Because NaOH and H_2SO_4, are a strong base and a strong acid, respectively, the dissociate completely to produce the following ions:

$$Na^+ + OH^- + 2\,H^+ + SO_4^- \rightarrow \text{salt} + H_2O$$

The H^+ and OH^- ions combine to form the H_2O on the product side. The Na+ and SO_4^- ions combine to form the salt sodium sulfate, or Na_2SO_4. The equation now reads:

$$NaOH + H_2SO_4 \rightarrow Na_2SO_4 + H_2O$$

The equation is unbalanced since the number of Na and H atoms on the reactant side and on the product side is unequal. The equation is balanced by putting a coefficient of 2 in front of the NaOH on the reactant side and a coefficient of 2 in front of the H_2O on the product side. The balanced equation is

$$2NaOH + H_2SO_4 \rightarrow Na_2SO_4 + 2H_2O$$

ANSWER KEY

Multiple Choice

1. b 2. a 3. d 4. d 5. c 6. a 7. c 8. b 9. b 10. d 11. d 12. a

True or False

1. True
2. True
3. False. Chemical bonds formed by van der Waals forces are much <u>weaker</u> than ionic, covalent, and metallic bonds.
4. False. Molecular crystals are characterized as being <u>soft</u> and having <u>low</u> melting points.
5. False. As a general rule, the solubilities of most solids <u>increase</u> with increasing temperature.
6. True
7. True
8. True
9. False. Pure acids are covalent compounds and form ions not by the separation of ions already present but by reacting with water.
10. False. The biochemical oxygen demand of an aquatic system is the amount of oxygen needed to completely oxidize the organic debris in a given volume of water.

Fill in the Blank

1. metallic
2. polar
3. nonpolar
4. saturated
5. ions
6. salinity
7. biochemical oxygen demand (or BOD)
8. strong
9. neutral
10. acidic, basic (or alkaline)

Matching

1. h 2. e 3. g 4. a 5. i 6. b 7. f 8. d 9. j 10. c

Chapter 11

CHEMICAL REACTIONS

OUTLINE

Quantitative Chemistry
11.1 Phlogiston
11.2 Oxygen
11.3 The Mole
11.4 Formula Units

Chemical Energy
11.5 Exothermic and Endothermic Reactions
11.6 Chemical Energy and Stability
11.7 Activation energy

Fuels
11.8 Liquid Fuels
11.9 Gas Fuels
11.10 Solid Fuels

Reaction Rates
11.11 Temperature
11.12 Other Factors
11.13 Chemical Equilibrium
11.14 Altering an Equilibrium

Oxidation and Reduction
11.15 Electrolysis
11.16 Electrochemical cells

GOALS

1. Discuss the phlogiston hypothesis and explain how Lavoisier's experiments showed it to be incorrect.
2. Define oxide.
3. Explain what is meant by the term mole.
4. Distinguish between exothermic and endothermic reactions.
5. Identify the nature of chemical energy.
6. Describe the relationship between the chemical energy absorbed or given off in a chemical change and the stabilities of the substances involved.
7. Explain what is meant by activation energy.
8. Compare gas, liquid, and solid fuels from the point of view of the pollution they produce when burned.
9. List the four factors that affect the speed of a chemical reaction.
10. Explain why reaction rates depend strongly on temperature.
11. Describe what is meant by a chemical equilibrium.
12. List the three main ways in which a chemical equilibrium can be altered to favor one direction over the other.
13. Distinguish between oxidation and reduction in terms of the electrons transferred in each case.
14. Describe electrolysis.
15. Explain the basic principle behind the operation of electrochemical cells.
16. Compare batteries and fuel cells.

CHAPTER SUMMARY

Utilization of energy sources, our technological advances, our standard of living, and life itself depend on the energy changes associated with chemical reactions. Chapter 11 discusses this important topic. The subject of quantitative chemistry is introduced, and the concept of the **mole** is presented. Chemical-energy changes are explained in terms of atomic structure, and chemical energy is defined. **Endothermic** and **exothermic** reactions are discussed, and factors that affect reaction rates are given. Reversible reactions and **chemical equilibrium** are introduced, and the methods for adjusting an equilibrium are discussed. **Oxidation** and **reduction** are defined, and examples of important **oxidation-reduction reactions** are given. A discussion of the environmental problems associated with the combustion of fuels and the methods used to alleviate these problems is presented.

CHAPTER OUTLINE

11-1. Phlogiston

A. Two early German chemists, Johann Becher and Georg Stahl, developed the **phlogiston** hypothesis.
B. According to the phlogiston hypothesis all substances that can be burned (undergo **combustion**) contain phlogiston. As burning takes place, the phlogiston leaves the substance only when air is present to absorb it.
C. The observation that when a metal is heated, the powder that forms weighs more than the original metal raised a problem with the phlogiston hypothesis. Since the powder should weigh less than the original metal according to the phlogiston hypothesis, it was assumed that phlogiston could sometimes have a negative mass.
D. The seventeenth century French chemist Antoine Lavoisier performed experiments that rejected the phlogiston hypothesis.
E. Lavoisier's experiments suggested that when tin is heated, the white powder formed results from the tin combining with a gas from the air. The increase in mass of the powder over the tin was the mass of the gas.
F. Lavoisier revolutionized chemistry by basing his ideas on accurate measurements.

11-2. Oxygen

A. The English chemist Joseph Priestley discovered a gas that Lavoisier later named **oxygen**.
B. When oxygen combines chemically with another substance, the process is called **oxidation**, and the substance is said to be **oxidized**.
C. An **oxide** is a compound of oxygen and another element.
    1. In general, the oxides of metals are solids.
    2. Oxides of other elements may be solid, liquid, or gaseous.

11.3 The Mole

A. A **mole** (mol) of any element is that amount of it whose mass in grams is equal to its atomic mass expressed in atomic mass units (u).

B. A mole of any element contains the same number of atoms as a mole of any other element.

C. The number of atoms per mole of any element is called **Avogadro's number** (N_0):

$$N_0 = 6.02 \times 10^{23} \text{ atoms/mol}$$

11.4 Formula Units

A. A formula unit for a chemical compound is the set of atoms given by its formula.

B. The formula mass of a substance is the sum of the atomic masses of the elements it contains, each multiplied by the number of times it appears in the formula of the substance.

C. Avogadro's number is equal to the number of formula units per mole of any substance.

D. A mole of any substance is that amount of it whose mass in grams is equal to its formula mass expressed in u.

E. A chemical equation can be interpreted in terms of moles as well as in terms of molecules or formula units.

11-5. Exothermic and Endothermic Reactions

A. Chemical changes that give off energy are called **exothermic reactions.**

B. Chemical changes that absorb energy are called **endothermic reactions.**

C. If the chemical reaction is exothermic, the reverse reaction will be endothermic. The amount of heat liberated by the exothermic reaction must be the same as the amount of heat absorbed by the endothermic reaction.

D. Dissociation is an endothermic process.

E. Neutralization is an exothermic process.

11-6. Chemical Energy and Stability

A. The more energy needed to decompose a substance, the greater the chemical stability of the substance (with a few exceptions).

B. Chemical energy is electron potential energy.

C. When electrons move to new locations during an exothermic reaction, some of their original potential energy is liberated.

   1. The freed energy may show itself in faster atomic or molecular motions that correspond to a higher temperature.
   2. The freed energy may excite outer electrons into higher energy levels from which they return by releasing photons of light.

D. In endothermic reactions, energy must be supplied to the atoms involved to enable some of their electrons to form bonds in which their potential energies are greater than before.

11-7. Activation Energy

A. In many exothermic reactions, the reactants must be **activated**, or supplied with sufficient outside energy to make the reaction start.
B. The energy needed to start a reaction is called the **activation energy**.
C. A molecule having enough energy to react is called an **activated molecule**.

11-8. Liquid Fuels

A. Petroleum, a mixture of various hydrocarbons, is the source of most liquid fuels.
B. The products of the incomplete combustion of gasoline, which can occur even under the best circumstances, may include carbon monoxide, unburned hydrocarbons, and nitrogen oxides, all of which are harmful.
C. There are three ways to reduce the amounts of pollutants in vehicle exhausts:
    1. Build an engine that completely burns a mixture of lead-free gasoline and air at the right temperature to minimize the emission of carbon monoxide, unburned hydrocarbons, and nitrogen oxides. Such an engine is difficult to produce.
    2. Use a **catalytic converter** to convert the polluting exhaust gases of burned lead-free gasoline into harmless gases.
    3. The use of "reformulated gasoline" which can reduce toxic emissions by 10 to 15 percent..

11-9. Gas Fuels

A. **Natural gas** is largely **methane,** CH_4.
B. An artificial gas fuel called **syngas** is produced by passing very hot steam over coal.
C. Underground coal can be gasified. Incomplete combustion of the coal will produce carbon monoxide, which can be used as a fuel or reacted with steam to give hydrogen.
E. Hydrogen gas is an excellent fuel produced from natural gas and by the electrolysis of water; however, large-scale hydrogen use is still years in the future.

11-10. Solid Fuels

A. Types of solid fuels include coal and wood.
B. There are three varieties of coal:
    1. Bituminous (soft) coal, the most common variety
    2. Anthracite (hard) coal
    3. Lignite
C. All varieties of coal contain sulfur, which plays a major role in the ecological problem of acid rain.

D. Upon combustion of coal, sulfur is oxidized to sulfur dioxide, SO_2. Nitrogen from the air is also oxidized to produce nitrogen oxides. These substances combine with atmospheric moisture to produce sulfuric acid and nitric acid, respectively.

E. The presence of these acids in the atmosphere has increased the average acidity of rainwater in much of Canada, the United States, and Europe.

F. The effects of acid rain include
   1. The dissolution and loss of plant nutrients from the soil.
   2. The conversion of nontoxic aluminum compounds in the soil to toxic compounds.
   3. The release of toxic metals into drinking water supplies.

G. One method of reducing acid rain is to "scrub" exhaust gases emitted by coal-fired power plants and metal refineries by mixing the gases with a lime ($CaCO_3$) solution.
   1. The sulfur dioxide in the exhaust gases combines with the lime to produce solid calcium sulfate.
   2. The resulting calcium sulfate sludge presents a disposal problem because of the vast amount produced.

11-11. Temperature

A. Reaction rates are increased by a rise in temperature.

B. A 10°C rise in temperature approximately doubles the speed of a chemical reaction occurring at or near room temperature.

C. Reaction rates increase with temperature primarily because the number of activated molecules increases.

D. Ionic reactions are very fast even at room temperatures because the ionic state itself is a form of activation.

11-12. Other Factors

A. Reaction rates of simple chemical reactions are proportional to the concentrations of the reactants.

B. Reaction speed depends on the amount of the solid surface exposed; the greater the surface area, the faster the reaction.

C. A **catalyst** is a substance that can change the rate of a chemical reaction without itself being permanently changed.

D. Catalysts accelerate reactions in different ways:
   1. The catalyst forms an unstable intermediate compound with one of the reactants which decomposes later in the reaction.
   2. The catalyst increases the reaction rate by producing activated molecules at its surface.

E. An **enzyme** is a protein catalyst that accelerates a specific chemical reaction within a living organism.

11-13. Chemical Equilibrium

A. Most chemical reactions are reversible.
B. In a **chemical equilibrium**, forward and reverse reactions occur at the same rate; the concentration of the reactants and products remain constant.
C. Many chemical reactions reach an intermediate equilibrium state instead of going to completion.

11-14. Altering an Equilibrium

A. In a chemical reaction that reaches equilibrium instead of going to completion, equilibrium conditions must be altered to increase the yield of a desired product.
B. There are three methods for adjusting a chemical equilibrium in one direction or the other:
   1. Change the concentration of one or more substances.
   2. Change the temperature.
   3. Change the pressure.

11-15. Electrolysis

A. Chemical reactions that involve electron transfer are called **oxidation-reduction reactions**.
B. **Oxidation** involves the loss of electrons by the atoms of the element in a chemical reaction; **reduction** involves the gain of electrons.
C. **Electrolysis**, in which free elements are liberated from a liquid by the passage of an electric current, is an example of an oxidation-reduction reaction.
D. In electrolysis, an **electrode** is a conductor through which electric current enters or leaves a solution.
E. Electrolysis is used in the process of **electroplating**, in which a thin layer of one metal is deposited on another metal.

11-16. Electrochemical Cells

A. Oxidation-reduction reactions can produce electric currents by having the transferred electrons pass through an external conductor as they go from one reactant to another.
B. **Electrochemical cells** use oxidation-reduction reactions to produce electric current.
C. Some examples of electrochemical cells include dry-cell batteries, lead-acid storage batteries, nickle-cadmium batteries, lithium ion batteries, and **fuel cells**.
D. Unlike dry-cell and lead-acid storage batteries, fuel cells can provide electric current indefinitely without having to be replaced or recharged, because the reactants are fed continuously.
   1. Fuel cells have been developed for use on spacecraft and to supply uninterruptible power for hospitals, banks, computer centers and similar type facilities.

2. Fuel cells utilizing a solid proton-exchange membrane (PEM)are being developed for use in automobiles and other vehicles.

KEY TERMS AND CONCEPTS

The questions in this section will help you to review the key terms and concepts from Chapter 11.

Multiple Choice

Circle the best answer for each of the following questions.

1. The scientist who discovered oxygen was
   a. Joseph Priestley
   b. Johann Becher
   c. Fritz Haber
   d. Antoine Lavoisier

2. According to the phlogiston hypothesis,
   a. all substances that can be burned contain phlogiston
   b. as a substance burns, the phlogiston it contains is turned into ash or powder
   c. metals are considered to be solid forms of pure phlogiston
   d. phlogiston must be present in air before a substance can be burned

3. The phlogiston hypothesis was overthrown as a result of the experiments conducted by
   a. Georg Stahl
   b. Joseph Priestley
   c. Antoine Lavoisier
   d. Johann Becher

4. When heated in the presence of air, tin turns into a white powder. The substance in the air that combines with the tin to form the white powder is
   a. nitrogen
   b. hydrogen
   c. carbon dioxide
   d. oxygen

5. In the preceding question, the tin has been
   a. activated
   b. oxidized
   c. tempered
   d. alloyed

6. The burning of coal, in which energy is liberated, is a(n)
   a. exothermic reaction
   b. endothermic reaction
   c. catalytic reaction
   d. equilibrium reaction

7. A compound is usually relatively stable if
   a. its decomposition is weakly exothermic
   b. it liberates energy when decomposed
   c. a great deal of energy is needed to decompose the compound
   d. the PE of its electrons is high

8. A good fuel is
   a. strongly exothermic
   b. cheap and abundant
   c. easy to transport
   d. all of the above

9. A catalytic converter works by
   a. trapping harmful gases so they are not part of the exhaust
   b. increasing the reaction temperature to increase the rate of combustion
   c. increasing the concentration of exhaust gas to promote more complete combustion
   d. promoting reactions that convert polluting gases into harmless ones by the actions of certain metals

10. Hydrogen is not widely used as a fuel at present because
   a. its combustion products are extremely "dirty"
   b. it is currently too expensive for ordinary purposes
   c. it has an extremely low heat of combustion
   d. it is scarce

11. A disadvantage of the use of syngas is
   a. the scarcity of the raw materials needed to produce syngas
   b. its high sulfur content
   c. problems related to transport and storage
   d. its carbon monoxide content

12. Which one of the following elements plays a major role in the formation of acid rain?
   a. carbon
   b. sulfur
   c. chlorine
   d. boron

13. The rate of a chemical reaction can be increased by
    a. lowering the temperature of the reaction
    b. reducing the concentration of the reactants
    c. increasing the surface area of a solid reactant
    d. lowering the activation energy

14. Catalysts used by living organisms to regulate their chemical processes are called
    a. hormones
    b. steroids
    c. enzymes
    d. lipids

15. A major disadvantage of fuel cells is their
    a. relatively short life span
    b. high weight in proportion to the electrical power they suppy
    c. lower efficiency as compared to current lead-acid storage batteries
    d. greater expense as compared to current lead-acid batteries

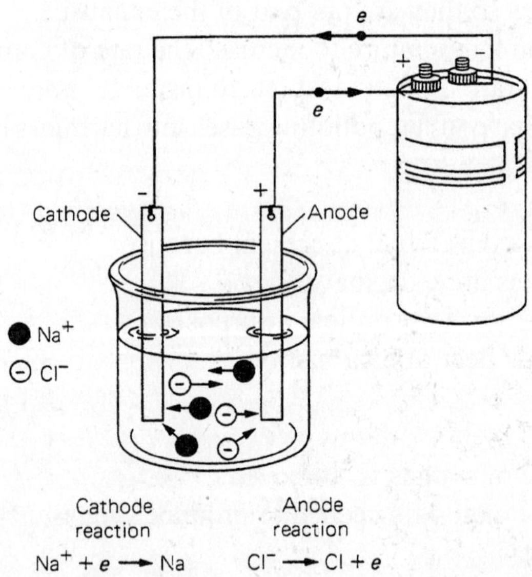

Refer to the above drawing to answer questions 16 through 18.

16. The drawing represents
    a. combustion of molten NaCl
    b. transmutation of NaCl
    c. catalytic dissociation of NaCl
    d. electrolysis of molten NaCl

17. The current in the molten NaCl consists of
    a. moving Na and Cl atoms
    b. moving Na^+ and Cl^- ions
    c. moving Cl^- ions only
    d. moving electrons

18. The process illustrated by the drawing is an example of
    a. an oxidation reaction only
    b. a reduction reaction only
    c. both oxidation and reduction reactions
    d. an acid-base neutralization reaction

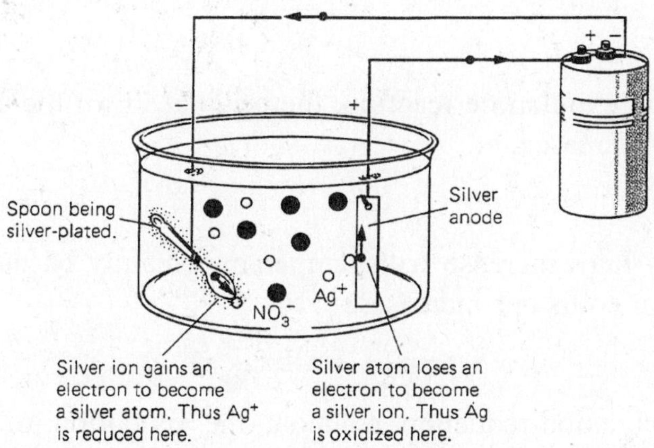

Refer to the above drawing to answer questions 19 through 20.

19. The nitrate ions (NO_3^-) remain in solution because
    a. nitrate ions are more soluble than silver ions
    b. silver atoms lose electrons at the anode more readily than nitrate ions
    c. the mass of a nitrate ion is less than that of a silver ion
    d. silver ions have a greater tendency to react with the metals than do nitrate ions

20. Silver atoms are reduced at the
    a. spoon
    b. anode
    c. battery
    d. cathode

True or False

Decide whether each statement is true or false. If false, either briefly state why it is false or correct the statement to make it true. See Chapter 1 or 2 for an example.

_________ 1. If a given chemical reaction is exothermic, the reverse reaction will be endothermic.

_________ 2. The amount of heat liberated by an exothermic reaction is the same as the amount of heat absorbed by the reverse endothermic reaction.

_________ 3. Chemical energy is really electron potential energy.

_________ 4. During an exothermic reaction, the original PE of the electrons of the reacting atoms is increased.

_________ 5. Reaction rates increase with temperature mainly because the total number of molecular collisions increases.

_________ 6. In an oxidation-reduction reaction, the oxidation of an element is always accompanied by the reduction of another.

_________ 7. Most acid rain is caused by the emissions of automobiles.

_________ 8. Using fossil fuels to produce hydrogen will help to reduce pollution and conserve resources.

Fill in the Blank

1. Rapid oxidation is called ____________________.

2. ____________________ number is equal to the number of atoms per mole of any element.

3. An ____________________ is a compound of oxygen and another element.

4. ____________________ chemical reactions liberate energy; ____________________ reactions absorb energy.

5. The energy needed to start a chemical reaction is called ____________________ energy.

6. ____________________, a mixture of various hydrocarbons, is the source of most liquid fuels.

7. A ____________________ is a substance that can change the rate of a chemical reaction without itself being permanently changed.

8. In an oxidation-reduction reaction, ____________________ involves the loss of electrons and ____________________ involves the gain of electrons.

9. In the process of ____________________, free elements are liberated from a liquid by the passage of an electric current.

10. Dry-cell batteries, lithium ion batteries, and fuel cells are all examples of ____________________ cells.

Matching

Match the name of the fuel on the left with its description on the right.

1. D gasoline
2. C natural gas
3. A syngas
4. H phlogiston
5. F wood
6. B bituminous coal
7. G anthracite coal
8. E lignite

a. made by passing very hot steam over coal
b. "soft" coal; the most common variety of coal
c. largely methane
d. derived from petroleum; contains octane
e. a very high sulfur coal
f. largely cellulose
g. "hard" coal
h. imaginary substance once thought to be present in all substances that burned

SOLVED PROBLEMS

Study the following solved example problems as they will provide insight into solving the problems listed at the end of Chapter11 in *The Physical Universe*. Review the mathematics refresher in the appendix of the text if you are unfamiliar with the basic mathematical operations presented in these examples.

Example 11-1

A beaker contains exactly 18 g of water. How many water molecules does that represent?

Solution

We know that one mole of water contains 6.02×10^{23} water molecules, so how many water molecules are there in 18 g of water?

Remember that a mole of any substance is that amount of it whose mass in grams is equal to its formula mass expressed in u. The formula mass of a molecule of water is the combined atomic masses of its hydrogen and oxygen atoms. According to the periodic table, hydrogen has an atomic mass of 1 u and oxygen has an atomic mass of 16 u. Since there are two hydrogen atoms and one oxygen atom in a molecule of water, the formula mass of a water molecule is

$$2 \times 1 \text{ u} + 1 \times 16 \text{ u} = 2 \text{ u} + 16 \text{ u} = 18 \text{ u}$$

Since a mole of any substance is that amount of it whose mass in grams is equal to its formula mass expressed in u, there are 6.02×10^{23} water molecules in 18 g of water.

Example 11-2

Zinc oxide is reduced by carbon to produce metallic zinc and carbon dioxide. How many moles of Zn will be produced if 24 g of C are used?

Solution

Begin by writing the reaction in the form of a chemical equation using chemical formulas. The equation is

$$ZnO + C \rightarrow Zn + CO_2$$

Check to see if the equation is balanced. There are one Zn atom and one C atom on each side of the equation; however, there is one O atom on the reactant side and two O atoms on the product side. The equation is unbalanced.

Begin by balancing the number of O atoms on both sides of the equation. A coefficient of 2 in front of the ZnO on the reactant side gives a total of two O atoms on both sides of the equation. The equation now reads:

$$2ZnO + C \rightarrow Zn + CO_2$$

Notice that the numbers of Zn atoms are still out of balance. A coefficient of 2 in front of the Zn on the product side balances the Zn atoms on both sides of the equation. The balanced equation is

$$2ZnO + C \rightarrow 2Zn + CO_2$$

To determine how many moles of Zn will be produced, remember that a chemical equation can be interpreted in terms of moles as well as in terms of molecules. the mole ratio of C to Zn in the balanced equation is 1:2.

This is a grams-to-moles problem. First we need to convert 24 grams of C to moles of C. The atomic mass of C is 12 u; therefore, one mole of C weighs 12 g. The number of moles in 24 g of C is

$$(24 \text{ g})\left(\frac{1 \text{ mole}}{12 \ g}\right) = 2 \text{ moles}$$

The mole ratio of Zn to C in the balanced equation is 2:1; therefore, the number of moles of Zn is

$$\text{Moles of Zn} = \left(\frac{2 \text{ moles of Zn}}{1 \text{ mole of C}}\right)(2 \text{ moles of C}) = 4 \text{ moles}$$

WEB LINKS

Investigate electrolysis at this interactive web site:

> http://www.chem.iastate.edu/group/Greenbowe/sections/projectfolder/flashfiles/electroChem/electrolysis10.html

For the latest information about the hydrogen economy, fuel cells, and related subjects visit:

> http://www.hydrogenhighway.com

ANSWER KEY

Multiple Choice

1. a 2. a 3. c 4. d 5. b 6. a 7. c 8. d 9. d 10. b 11. d 12. b 13. c 14. c 15. d 16. d 17. b 18. c 19. b 20. b

True or False

1. True
2. True
3. True
4. False. During an exothermic reaction, the original PE of the electrons of the reacting atoms is liberated and thereby reduced.
5. False. Reaction rates increase with the temperatures mainly because the number of activated molecules increases.
6. True
7. False. Most acid rain is caused by the emissions from coal-burning power plants and metal refineries.
8. False. Using fossil fuels to produce hydrogen will do noting to reduce pollution and conserve resources.

Fill in the Blank

1. combustion
2. Avogadro's
3. oxide
4. exothermic, endothermic
5. activation
6. petroleum
7. catalyst
8. oxidation, reduction
9. electrolysis
10. electrochemical

Matching

1. d 2. c 3. a 4. h 5. f 6. b 7. g 8. e

Chapter 12

ORGANIC CHEMISTRY

OUTLINE

GOALS

1. Define organic chemistry.
2. Discuss the covalent bonding behavior of carbon atoms.
3. Define hydrocarbon and alkane and explain why the alkane series of hydrocarbons is so important.
4. Describe how fractional distillation works.
5. Compare the ways in which cracking and polymerization increase the yield of gasoline from petroleum.
6. Discuss the properties of alkane molecules that make oil spills so serious.
7. Compare molecular and structural formulas and explain why the latter are so useful in organic chemistry.
8. Define isomer.
9. Compare saturated and unsaturated compounds and explain why the latter are more reactive.
10. Draw the structural formula of benzene.
11. Explain what is meant by a functional group and list several important functional groups.
12. Distinguish between monomers and polymers and list several examples of polymers.
13. Explain why nylon is called a polyamide and dacron a polyester.
14. Identify carbohydrates and discuss what they are used for by living things.
15. Describe photosynthesis and give the reasons for its importance.
16. Identify lipids and discuss what they are used for by living things.
17. Identify cholesterol and discuss its role in heart disease.
18. Identify proteins and discuss what they are used for by living things.
19. Account for the wide variety of proteins.
20. Explain the importance of soil nitrogen and list the ways in which it is replenished.
21. Describe the structure of the nucleic acid DNA and list the three fundamental attributes of life it is responsible for.
22. Define gene, genome, and chromosome.

CHAPTER SUMMARY

Chapter 12 introduces the important branch of chemistry known as **organic chemistry**, the chemistry of carbon compounds. The general properties of carbon compounds and the use of **structural formulas** in the representation of organic compounds are presented. The distinction is made between **saturated** and **unsaturated** compounds and between **aromatic** and **aliphatic** compounds. The presence of **functional groups** and the behavior of representative organic compounds are discussed. The classes of organic compounds found in living organisms are presented, and the equations for the oxidation of glucose and the process of **photosynthesis** are given. The molecular basis of inheritance and the origin of life are examined.

CHAPTER OUTLINE

12-1. Carbon Bonds

A. **Organic chemistry** is the chemistry of carbon compounds; **inorganic chemistry** is the chemistry of compounds of all elements other than carbon.

B. Carbon atoms form four covalent bonds with each other as well as with other atoms.

C. The general properties of carbon compounds are:

   1. Most carbon compounds are nonelectrolytes.
   2. The reaction rates of carbon compounds are usually slow.
   3. Many carbon compounds oxidize slowly in air but rapidly if heated.
   4. Most carbon compounds are unstable at high temperatures.

12-2. Alkanes

A. The simplest carbon compounds, **hydrocarbons**, contain only carbon and hydrogen.

B. **Alkanes** are hydrocarbons whose molecules have only single carbon-carbon bonds.

C. Examples of alkanes include methane (CH_4), propane (C_3H_8), and butane (C_4H_{10}).

D. Petroleum and natural gas consist largely of alkanes.

12-3. Petroleum Products

A. There are three processes by which petroleum products are derived:

   1. **Fractional distillation**, in which crude oil is heated and its vapors are collected and condensed at progressively higher temperatures.
   2. Catalytic cracking, in which heavy hydrocarbons are **cracked** into smaller molecules by heating them under pressure in the presence of catalysts.
   3. **Polymerization**, in which the molecules of lighter hydrocarbons are polymerized by joining the smaller molecules into larger ones under the influence of heat, pressure, and catalysts.

B. Alkane molecules are nonpolar, insoluble in water, and rather chemically unreactive. These properties, plus their toxicity to living organisms, makes the discharge of petroleum and petroleum products into the sea a serious environmental matter.

12-4. Structural Formulas

A. **Structural formulas** represent the covalent bonds between the atoms of a molecule by dashes and show how the atoms are linked together. Each dash stands for a pair of shared electrons.

B. The number of bonds an atom forms in an organic compound is the same as the number of electrons it has to gain or lose to achieve a closed outer shell.

12-5. Isomers

A. **Isomers** are compounds that have the same molecular formulas but different structural formulas and properties.

B. Flipping a structural formula end-over-end does not give the formula of an isomer.

12-6. Unsaturated Hydrocarbons

A. The linkage of two adjacent carbon atoms in a carbon compound by the sharing of two electron pairs (two covalent bonds) is called a **double bond**.

B. Three electron pairs (three covalent bonds) are shared by adjacent carbon atoms in a **triple bond**.

C. **Unsaturated compounds** have double or triple carbon-carbon bonds and are more reactive than **saturated compounds**, which have only single carbon-carbon bonds (alkanes and similar compounds).

12-7. Benzene

A. **Benzene**, C_6H_6, is a clear liquid that is insoluble in water, has a strong odor, and has its six C atoms arranged in a flat hexagonal ring.

B. In benzene, six electrons are shared by the six carbon atoms in the ring and belong to the molecule as a whole and not to any particular pair of atoms. These electrons are **delocalized**.

C. **Aromatic** compounds are organic compounds that contain one or more benzene rings in their molecules.

D. Many aromatic compounds, such as toluene, have strong odors.

E. **Aliphatic** compounds are organic compounds that do not contain benzene rings.

12-8. Hydrocarbon Groups

A. Hydrocarbon **derivatives** are compounds obtained by substituting other atoms or atom groups for one or more of the H atoms in hydrocarbon molecules.

B. The carbon-hydrogen atom groups that appear in the hydrocarbon derivatives are named from the parent hydrocarbons.

Examples:

Parent hydrocarbon	Name of group
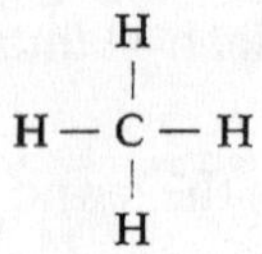 Methane	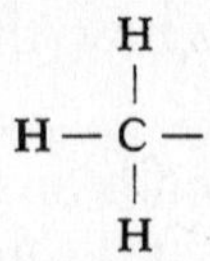 Methyl group
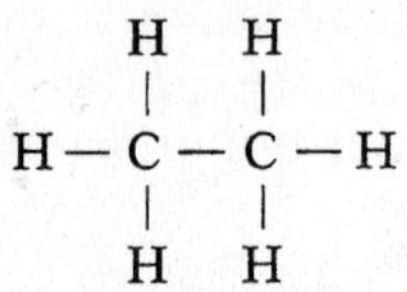 Ethane	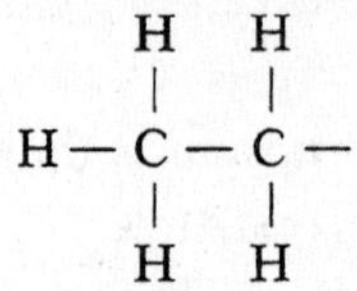 Ethyl group
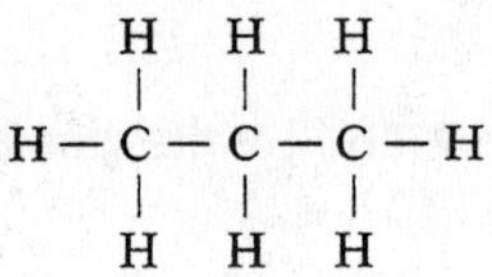 Propane	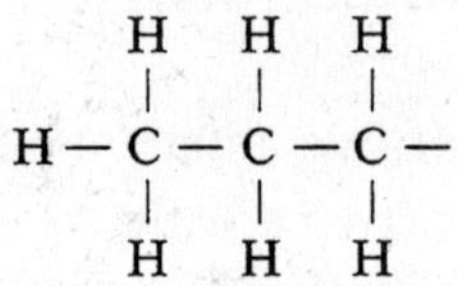 Propyl group

12-9. Functional Groups

A. A **functional group** is a group of atoms whose presence in an organic molecule largely determines its chemical behavior.

B. Some classes of organic compounds and their functional groups are:

1. **Alcohols**. Alcohols contain the hydroxyl group (—OH). Examples include ethanol, ethylene glycol, and glycerol. Ethanol in beverages is produced by fermentation of sugar by yeast enzymes.
2. **Ethers**. An ether has an oxygen atom bonded between two carbon atoms (– O –); ethers are widely used solvents.
3. **Aldehydes** and **ketones**. Aldehydes and ketones both contain the carbonyl atom group $-\overset{O}{\overset{\|}{C}}-$. In aldehydes, the carbonyl group is at the end of the molecule with a hydrogen atom attached to the carbon atom. In ketones, the carbonyl group is inside a moleculebetween adjacent carbon atoms. An example of an aldehyde is formaldehyde; an example of a ketone is acetone.

4. **Organic acids**. Organic acids contain the carboxyl group (—COOH). Examples include formic acid, acetic acid, and citric acid.

5. **Esters**. Esters contain the ester group 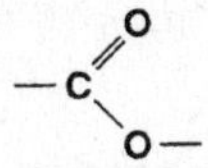 . Many esters have fruity or flowerlike odors. Examples include ethyl acetate, nitroglycerin, and animal and vegetable fats.

12-10. Polymers

A. A **polymer** is a long chain of simple molecules (**monomers**) linked together.
B. Polymers that contain the **vinyl group** are classed as vinyls.

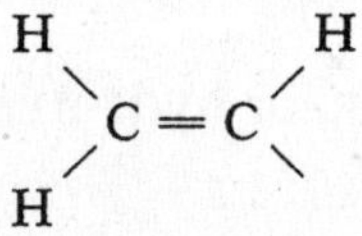

Vinyl group

C. Some examples of polymers include Styrofoam, Teflon, Orlon, and Plexiglas (or Lucite). Plexiglas is **thermoplastic**, meaning it softens and can be shaped when heated but becomes rigid again on cooling.
D. A **copolymer** is a polymer that consists of two different monomers. Dynel and Saran Wrap are examples.
E. Certain monomers that contain two double bonds in each molecule form flexible, elastic polymers called **elastomers**; rubber and neoprene are examples.
F. **Polyamides** and **polyesters** are polymers produced by chemical reactions rather than by the polymerization of monomers.
   1. Polyamides contain amide groups linking the structural elements. The synthetic fiber nylon is an example.

Amide linkage

   2. Polyesters contain ester groups linking their structural elements. The synthetic fiber Dacron is an example.

12-11. Carbohydrates

A. **Carbohydrates** are compounds of carbon, hydrogen, and oxygen atoms whose ratio between hydrogen atoms and oxygen atoms is 2 to 1. Examples include sugars and starches.
   1. **Monosaccharides** are simple sugars. Glucose, fructose, galactose, and mannose are examples.
   2. **Disaccharides** are composed of two simple sugar monomers per molecule. Examples are sucrose, lactose, and maltose.
   3. **Polysaccharides** are complex sugars that consist of chains of more than two simple sugars. Examples include **cellulose**, **starch**, **chitin**, and **glycogen**.

B. The oxidation of glucose provide plants and animals with energy. The overall reaction is:

$$C_6H_{12}O_6 + 6O_2 \rightarrow 6CO_2 + 6H_2O + \text{energy}$$

C. Carbohydrates are broken down, or **hydrolyzed**, into simple sugars during digestion.

12-12. Photosynthesis

A. **Photosynthesis** is the reverse process of the oxidation of glucose. The overall reaction is

$$6CO_2 + 6H_2O + \text{energy} \rightarrow C_6H_{12}O_6 + 6O_2$$

B. **Chlorophyll** is the catalyst for photosynthesis. The energy comes from sunlight.

C. Atmospheric oxygen comes from photosynthesis.

12-13. Lipids

A. **Lipids** are organic compounds composed of the elements carbon, oxygen, and hydrogen, but contain less O relative to C and H than do carbohydrates. Examples include fats, oils, waxes, and sterols.

B. A fat molecule consists of a glycerol molecule with three **fatty acid** molecules attached.
   1. Saturated fats are solids at room temperatures; unsaturated fats are liquids.
   2. Fats provide nearly twice as much energy per gram than do carbohydrates.
   3. **Hydrogenation** is used to convert liquid fat into solid fat to make margarine and other products. During hydrogenation, H atoms are added to the double-bonded C atoms in a liquid fat, which produces a solid fat.

C. **Cholesterol** is a lipid found in the bloodstream. Deposits of cholesterol within arteries can cause the serious condition **atherosclerosis**.

12-14. Proteins

A. **Proteins** are compounds of carbon, hydrogen, oxygen, nitrogen, and often sulfur and phosphorus. They are the chief constituents of living cells.
B. Proteins are composed of 20 different **amino acids** linked by **peptide bonds** to form **polypeptide chains** that are usually coiled or folded in intricate patterns.
C. The shape of a protein molecule and its sequence of amino acids give it a unique character that is reflected in its biological activity.
D. The human body can synthesize only some of the 20 amino acids needed to manufacture the proteins essential to life. The amino acids the body cannot synthesize must be provided in the diet.

12-15 Soil Nitrogen

A. Nitrogen compounds in the soil are removed by plants to supply nitrogen for the formation of plant proteins.
B. The decay of dead plants, animals, and animal wastes returns nitrogen to the soil but does not completely replenish what has been removed by plants.
C. Soil nitrogen is returned to the soil by two means:
    1. "Nitrogen-fixing" bacteria convert atmospheric nitrogen into nitrogen compounds.
    2. Lightning causes atmospheric nitrogen to combine with oxygen to form nitrogen oxides. These are returned to the earth in rainwater.
D. Man has disturbed the natural nitrogen cycle, resulting in a number of environmental problems.

12-16. Nucleic Acids

A. **Nucleic acids** consist of long chains of units called **nucleotides**.
B. Each nucleotide has three parts:
    1. A **phosphate group** (PO_4)
    2. A **pentose sugar**: **ribose** ($C_5H_{10}O_5$) in **ribonucleic acid** (RNA) and **deoxyribose** ($C_5H_{10}O_4$) in **deoxyribonucleic acid** (DNA)
    3. A **nitrogen base**
        a. The nitrogen bases in DNA are adenine, guanine, cytosine, and thymine.
        b. The nitrogen bases in RNA are the same as in DNA, except uracil replaces thymine.
C. DNA has the form of a double helix linked by the nitrogen bases; the sequence of these bases is the genetic code.
    1. **Chromosomes** consist of DNA molecules.
    2. DNA molecules govern protein synthesis in cells and, by making copies of themselves, can pass on the genetic code to new cells, thus ensuring that new cells have the same characteristics, or **heredity,** as the original cell.

3. Changes in the sequence of the bases in a DNA molecule can result in a **mutation**.

D. RNA differs from DNA in the following ways:
   1. The sugar in RNA is ribose, not deoxyribose as in DNA.
   2. RNA molecules are much smaller then DNA molecules.
   3. RNA molecules usually consist of only single strands of nucleotides.

E. The set of coded DNA instructions for each protein is called a **gene**, and the entire collection of genes is called the **genome** of an organism.

F. James D. Watson and Francis H. C. Crick discovered the structure of DNA in 1953.

12-17. Origin of Life

A. Life probably originated in or near the early oceans when the composition of the atmosphere likely consisted of water, methane, ammonia, carbon dioxide, and hydrogen cyanide.

B. A certain portion of these gases were dissolved in the early oceans to form a sort of "chemical soup."

C. Of the reactions that took place in this chemical soup, five classes of important biological molecules likely formed. These are the sugars, glycerol, the fatty acids, the amino acids, and the nitrogen bases.

E. Reactions between these molecules likely lead to more complex compounds and ultimately to the first molecules able to replicate themselves. These eventually gave rise to the first primitive cells.

KEY TERMS AND CONCEPTS

The questions in this section will help you review the key terms and concepts from Chapter 12.

Short Answer

In the space provided, write the name of the functional group that appears in each of the structural formulas.

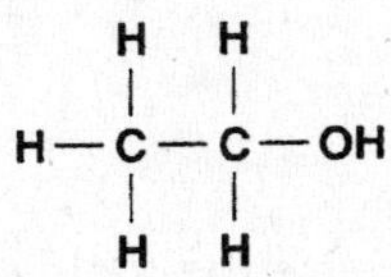

1. ______________________

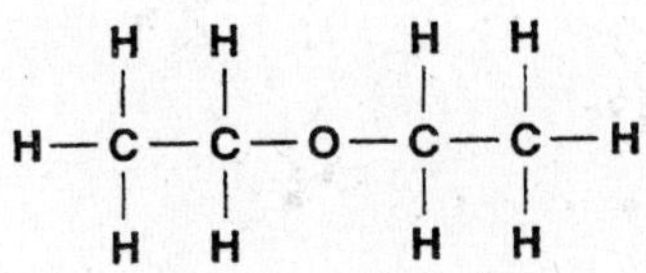

2. ______________________

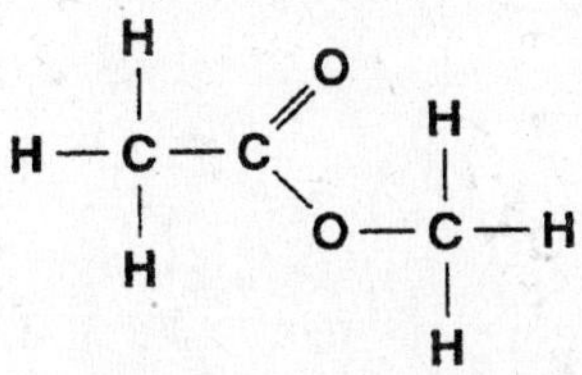

3. ______________________

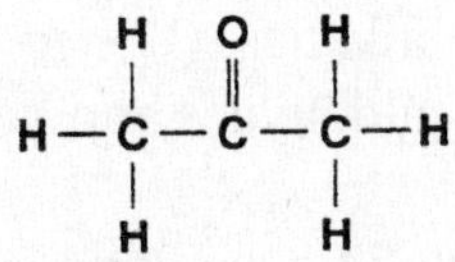

4. ______________________

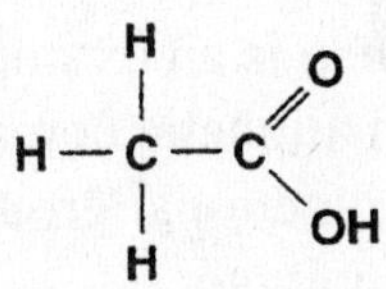

5. ______________________

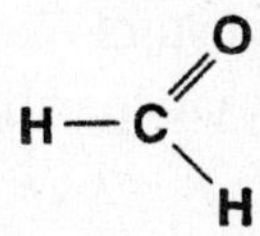

6. ______________________

Multiple Choice

Circle the best answer for each question.

1. Carbon compounds
   a. are mostly electrolytes
   b. usually have slow reaction rates
   c. rapidly oxidize in air
   d. tend to be stable at high temperatures

2. The simplest organic compounds are the
   a. hydrocarbons
   b. lipids
   c. carbohydrates
   d. monosaccharides

3. Natural gas and petroleum consist mainly of
   a. unsaturated hydrocarbons
   b. aromatic hydrocarbons
   c. alkanes
   d. aldehydes and ketones

4. Catalytic cracking and polymerization are two methods used to increase the yield of
   a. naphtha
   b. pentane
   c. gasoline
   d. alcohol

5. Benzene is an example of
   a. an aliphatic compound
   b. a functional group
   c. a polymer
   d. an aromatic compound

6. Vinyls, Styrofoam, Plexiglas, and Teflon are examples of
   a. monomers
   b. polymers
   c. polyesters
   d. polysaccharides

7. Sugars and starches are examples of
   a. lipids
   b. carbohydrates
   c. proteins
   d. nucleic acids

8. The catalyst for photosynthesis is
   a. DNA
   b. cholesterol
   c. cellulose
   d. chlorophyll

9. Fat molecules consist of three fatty acid molecules attached to
   a. a benzene ring
   b. an amino acid
   c. an alcohol molecule
   d. a glycerol molecule

10. The organic molecule that can replicate itself is
   a. DNA
   b. an amino acid
   c. acetaldehyde
   d. ethylene glycol

True or False

Decide whether each statement is true or false. If false, either briefly state why it is false or correct the statement to make it true. See Chapter 1 or 2 for an example.

__________ 1. Carbon compounds are produced only by plants and animals.

__________ 2. A structural formula shows how many atoms of each kind are present and how these atoms are linked together.

__________ 3. The number of bonds an atom forms in an organic compound is the same as the number of electrons it has to gain or lose to achieve a closed outer shell.

__________ 4. Isomers of compounds having the same molecular formula also have the same set of physical and chemical properties.

__________ 5. The process of the oxidation of glucose is the reverse of photosynthesis.

__________ 6. Unsaturated fats are normally solids at room temperature.

__________ 7. Proteins consist of long chains of simple sugar units.

__________ 8. The human body can synthesize all of the 20 amino acids required to make human proteins.

__________ 9. The nitrogen bases in RNA are uracil, adenine, guanine, and cytosine.

__________ 10. The sequence of nitrogen bases in a DNA molecule represents the genetic code for a specific organism.

Fill in the Blank

1. ____________________ chemistry is the chemistry of carbon compounds.

2. ____________________ are the simplest group of carbon compounds because they contain only carbon and hydrogen.

3. ____________________ are organic compounds that have the same molecular formulas but different structural formulas and properties.

4. ____________________ compounds are able to add other atoms to their molecules; ____________________ compounds cannot add other atoms to their molecules.

5. ____________________ compounds are organic compounds that contain one or more benzene rings; ____________________ compounds are organic compounds that do not contain benzene rings.

6. Alcohols react with acids to form ____________________.

7. A ____________________ group is a group of atoms whose presence in an organic compound largely determines its chemical behavior.

8. A ____________________ is a long chain of simple molecules chemically linked together.

9. ____________________ material softens and can be shaped when heated but becomes rigid again upon cooling.

10. A ____________________ is a compound of carbon, hydrogen, and oxygen whose ratio of hydrogen to oxygen is 2:1.

11. ____________________ are biological molecules that include fats, oils, and waxes.

12. In a ____________________, amino acids are linked by peptide bonds.

13. Nucleic acids consist of long chains of units called ____________________.

14. ____________________ controls the development and functioning of cells by determining the proteins cells make.

15. The set of instructions for each protein is called a ____________________.

Matching

Match the carbon compound on the right with its correct description or use.

1._____ acetone

2._____ nitrile rubber

3._____ butane

4._____ acetic acid

5._____ Teflon

6._____ Spectra

7._____ acetylene

8._____ cellulose

9._____ chitin

10._____ cholesterol

a. used in welding and metal-cutting torches
b. component of the shells of crustaceans
c. lipid associated with atherosclerosis
d. chief component of plant cell walls
e. used as fuel in stoves and furnaces
f. gives vinegar its characteristic taste
g. strongest synthetic fiber
h. solvent used in paints and nail polish
i. an elastomer used to line gasoline hoses
j. used for industrial seals and bearings and as a nonstick coating for cookware

SOLVED PROBLEMS

Study the following solved example problem as it will provide insight into solving the problems listed at the end of Chapter12 in *The Physical Universe*.

Example 12-1

Heating ethanol in the presence of concentrated sulfuric acid (which acts as a catalyst) yields the commercial solvent diethyl ether ($C_4H_{10}O$) and water. What is the structural formula of diethyl ether?

Solution

The name diethyl ether indicates that in each molecule of the substance there are two ethyl groups and one ether functional group (– O –). There can be only one ether functional group because there is only one oxygen atom in the formula for diethyl ether, and that single oxygen atom must be part of the ether functional group.

Recall that the ether functional group consists of a oxygen atom bonded between two carbon atoms, so part of the structural formula for diethyl ether must be

Each carbon atom must belong to an ethyl group. If the two ethyl groups are represented, the structural formula is

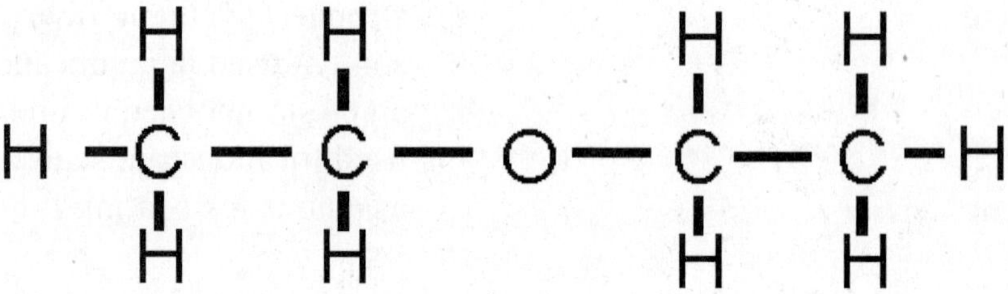

Note that in the structural formula for diethyl ether there are four carbon atoms, ten hydrogen atoms, and one oxygen atom; therefore, the numbers of the atoms of each element are the same as those in the molecular formula. Also note that the bonding capacities for each atom are correct. Each carbon atom participates in four bonds, each hydrogen atom in one, and the oxygen atom in two.

WEB LINKS

View interactive organic chemistry movies on such topics as representing compounds, aldehydes and ketones, stereochemistry, and others at

ehttp://www.colby.edu/chemistry/OChem/demoindex.html

Watch the construction of a DNA molecule at

http://www.pwc.k12.nf.ca/wadey/biotech/dna1.swf

ANSWER KEY

Short Answer

1. hydroxyl 2. ether 3. ester 4. ketone 5. carboxyl 6. aldehyde

Multiple Choice

1. b 2. a 3. c 4. c 5. d 6. b 7. b 8. d 9. d 10. a

True or False

1. False. It was once believed that carbon compounds could be produced only by living organisms. This idea was disproved by the German chemist Friedreich Wohler when he prepared the organic compound urea in his laboratory.
2. True
3. True
4. False. Isomers have different properties, both physical and chemical.
5. True
6. False. Most unsaturated fats are liquids at room temperature.
7. False. Proteins consist of long chains of amino acids.
8. False. The human body can synthesize only some of the 20 amino acids needed to make human proteins.
9. True
10. True

Fill in the Blank

1. organic
2. hydrocarbons
3. isomers
4. unsaturated; saturated
5. aromatic, aliphatic
6. esters
7. functional
8. polymer
9. thermoplastic
10. carbohydrate
11. lipids
12. protein
13. nucleotides
14. DNA
15. gene

Matching

1. h 2. i 3. e 4. f 5. j 6. g 7. a 8. d 9. b 10. c

Chapter 13

ATMOSPHERE AND HYDROSPHERE

OUTLINE

GOALS

1. List in order of abundance the four chief ingredients of dry air near ground level.
2. Distinguish between the troposphere, stratosphere, mesosphere, and thermosphere.
3. Define ozone and explain why the ozone layer in the upper atmosphere is so important.
4. List some of the uses of chlorofluorocarbons (CFCs) and describe their effect on the ozone layer.
5. Define what is meant by saturated air.
6. List the ways in which clouds come into being.
7. Describe what causes rain and snow to fall from a cloud.
8. Describe the difference between sleet and hail.
9. Define insolation and describe the greenhouse effect.
10. Discuss why temperatures vary around the earth.
11. Explain how the seasons of the year originate.
12. Describe what is meant by a convection current.
13. State the influence of the coriolis effect on wind direction in the northern and southern hemispheres.
14. Sketch on a map the main surface wind systems of the world and name them and the belts of relative calm that separate them.
15. Explain what a jet stream is.
16. Compare cyclones and anticyclones and describe the motion of air in each of them.
17. Compare warm and cold fronts and describe what happens when a cold front overtakes a warm front.
18. Distinguish between weather and climate.
19. Describe the ice ages and the variations in the earth's motions that may be responsible for them.
20. Discuss the relationship between the carbon dioxide content of the atmosphere and global warming.
21. List the ways in which the oceans affect climates.

CHAPTER SUMMARY

Chapter 13 surveys the earth's **atmosphere** and **hydrosphere**. The composition and the layers of the atmosphere are described, and atmospheric moisture and cloud formation are discussed. The atmosphere's important role in maintaining the earth's heat balance is presented. The origin of the winds and the general circulation of the atmosphere are described. Weather systems associated with **air masses** are discussed, and the characteristics of **cyclones** and **anticyclones** are given. The climates of tropical, middle-latitude, and polar regions are described, and the origins of climatic change are examined. The nature of the world's ocean basins is presented, and the role of the oceans in moderating climate is discussed.

CHAPTER OUTLINE

13-1. Regions of the Atmosphere

A. The **atmosphere** is the envelope of air surrounding the earth and consists mainly of nitrogen (78 percent) and oxygen (21 percent).
B. The **hydrosphere** includes all the water of the earth's surface.
C. The **troposphere** is the lowest and the densest layer of the atmosphere.
   1. Temperature decreases with height in the troposphere.
   2. All weather occurs in the troposphere.
D. The **stratosphere** is the layer of the atmosphere above the troposphere.
   1. **Ozone** (O_3) in the stratosphere absorbs harmful solar ultraviolet radiation.
   2. Certain pollutants, especially the gaseous **chlorofluorocarbons** (CFCs), cause the breakdown of O_3 molecules, thus allowing additional ultraviolet radiation to reach the earth's surface.
E. The portion of the atmosphere above the stratosphere extending from 50 to 80 km above the earth is called the **mesosphere**. The mesosphere is characterized by decreasing temperature with altitude, changing from about 10°C at 50 km to about -75°C at 80 km.
F. Above the mesosphere, extending from 80 km to about 600 km, is the **thermosphere**.
   1. Temperature increases within the thermosphere to about 2000°C, although the density of the thermosphere is so low that there is very little total heat energy present.
   2. An ionized region within the thermosphere called the ionosphere reflects radio waves back to earth and makes possible long-distance radio communication.

13-2. Atmospheric Moisture

A. **Water vapor** consists of water molecules that have escaped, or **evaporated**, from a body of water at a temperature below the boiling point of water.
B. **Humidity** refers to the water vapor content of air.
C. Air is said to be **saturated** when it holds the maximum amount of water vapor at a given temperature.

13-3. Clouds

A. When air that contains water vapor is cooled past its saturation point, some of the vapor condenses to form dew, fog, or clouds.
B. Rising, warm, moist air expands and cools as atmospheric pressure decreases. This causes water vapor to condense into clouds consisting of ice crystals or water droplets, depending on temperature.
C. Condensation nuclei, such as airborne salt and dust particles, must be present before condensation can occur.
D. Three processes in the atmosphere can cause cloud formation:
    1. A warm air mass moving horizontally meets a land barrier such as a mountain and rises.
    2. An air mass is heated by contact with a warm part of the earth's surface and rises by **convection**.
    3. A warm air mass meets a cooler air mass and, being less dense, is forced upward over the cooler mass.
E. Rapid cooling of a cloud leads to rapid condensation and the precipitation of rain or snow, depending on temperature.
F. **Sleet** consists of raindrops that freeze as they fall; **hail** forms as violent up- and downdrafts associated with thunderstorms cause frozen raindrops to accumulate layers of ice as they alternately rise and fall.
G. Clouds are sometimes **seeded** with condensation nuclei of silver iodide crystals or "dry ice" (solid CO_2) to induce precipitation.

13-4. Atmospheric Energy

A. **Meteorology** is the study of weather.
B. **Weather** refers to the temperature, humidity, air pressure, cloudiness, and precipitation at any given time at a given place.
C. **Climate** is a summary of weather conditions over a period of years over a large area, including how temperatures and rainfall vary with the seasons.
D. Solar energy arriving at the upper atmosphere is called **insolation** and is mainly in the form of visible light.
E. Insolation reaching the earth's surface is absorbed and becomes heat. This heat energy is radiated back into the atmosphere as long-wavelength infrared radiation.
F. This long-wavelength infrared radiation is readily absorbed by atmospheric CO_2 and water vapor, thus preventing its rapid escape into space. This heating of the atmosphere is called the **greenhouse effect**.
G. Although temperatures vary around the earth, the earth's average temperature changes very little with time. There is a balance between incoming radiation and outgoing energy.
H. Winds and the ocean currents carry energy from tropical regions to the cooler polar regions.

13-5. The Seasons

A. The earth is nearest the sun in January and is farthest from the sun in July.
B. The earth's seasons are caused by the tilt of its axis together with its annual orbit around the sun.
C. Because of the tilt of the earth's axis, in half of the year one hemisphere receives more direct sunlight than the other hemisphere, and in the other half of the year it receives less.
D. On about June 22 the noon sun is at its highest in the sky in the northern hemisphere, and the period of daylight is longest.
F. The **solstices** occur on about December 22 the noon sun is at its lowest in the sky, and the period of daylight is shortest.
G. The **equinoxes** occur on about March 21 and on about September 23 the noon sun is directly overhead at the equator and the periods of daylight and darkness are equal.

13-6. Winds

A. **Winds** are horizontal movements of air that take place in response to the pressure differences in the atmosphere. The greater the pressure difference, the faster the wind.
B. Uneven heating of the earth's surface creates **convection currents** in which rising warm air is cooled and then moves downward toward the earth.
C. The **coriolis effect** is the deflection of moving things such as air currents due to the earth's rotation. The deflection is to the right in the northern hemisphere and to the left in the southern hemisphere.
D. Air rushing into a low-pressure region follows a counterclockwise spiral path in the northern hemisphere and a clockwise spiral path in the southern hemisphere.

13-7. General Circulation of the Atmosphere

A. The north and the south winds coming from the polar regions are deflected by the coriolis effect into large-scale eddies that lead to a generally eastward drift in the middle of each hemisphere and a westward drift in the tropics.
B. The steady easterly winds on each side of the equator are called the **trade winds**.
C. The region of light, erratic wind along the equator, where the principal movement of air is upward, constitutes the **doldrums**.

D. The largely calm belts that separate the trade winds in both hemispheres from the prevailing westerlies poleward of them are called the **horse latitudes**.
E. The **jet streams** are narrow, meandering, high-speed winds in the upper troposphere.

13-8. Middle-Latitude Weather Systems

A. The movement of warm and cold air masses through the belts of the westerlies determines the weather in North America.

B. An **air mass** is a large body of air that is relatively uniform in temperature and moisture content.

C. A **cyclone** is a weather system centered on a low-pressure region.
   1. Cyclonic winds blow in a counterclockwise spiral in the northern hemisphere and in a clockwise spiral in the southern hemisphere.
   2. Cyclones usually bring unstable weather conditions with clouds, rain, strong winds, and abrupt temperature changes.

D. An **anticyclone** is a weather system centered on a high-pressure region.
   1. Anticyclonic winds blow in a clockwise spiral in the northern hemisphere and in a counterclockwise spiral in the southern hemisphere.
   2. The weather associated with anticyclones is usually settled with clear skies and little wind.

E. Middle-latitude cyclones originate at the **polar front**, the boundary between the cold polar air mass and the warm maritime tropical air mass.

F. A **warm front** forms when a relatively warm air mass moves to a region occupied by colder air. A prolonged period of cloudiness and precipitation often follows a warm front.

G. A **cold front** forms when a cold air mass moves into a region occupied by warmer air. Rainfall, if it occurs, is heavier and of shorter duration than along a warm front.

H. An **occluded front** forms when a cold front overtakes a warm front and lifts the warm air off the ground.

13-9. Tropical Climates

A. Weather in the doldrums is hot with almost daily rains and light, changeable winds.

B. Weather in the horse latitudes is dry with light winds or calms.

C. Weather in the trade wind belts is for the most part dry, although seasonal shifts of the trade wind belts give rainfall during part of the year along their equatorial margins.

13-10. Middle-Latitude Climates

A. The belts of prevailing westerlies generally have moderate average temperatures, although continental interiors show pronounced seasonal variations.

B. In the United States, mountain ranges along the West Coast wring out moisture from the prevailing westerlies so that their western slopes receive abundant rainfall. Their eastern slopes and the continental interior are dry. Rainfall increases eastward across the country.

C. Polar weather is characterized by short summers and long cold winters.

13-11. Climatic Change

A. Evidence indicates that climates are subject to long-term change.

B. The **ice ages** were periods of extreme climatic change, the last one reaching its peak about 21,000 years ago.

C. The "Little Ice Age" was a period of cool summers and very cold winters that took place during the first half of the seventeenth century.
D. The late nineteenth century experienced a trend toward a warmer worldwide climate that peaked about 1940. Recently, global temperatures have again started upward.
E. Global warming has caused a rise in sea level, mostly due to the thermal expansion of sea water.
F. Continued melting of the earth's ice sheets and glaciers contributes to rising sea level and other environmental changes.

13-12. Origins of Climatic Change

A. Fluctuations in the sun's energy output may account for some aspects of climatic change.
B. The Milankovitch theory of climatic change relates large-scale climatic changes to periodic changes that occur in the tilt of the earth's axis, the shape of earth's orbit, and the time of year when earth is closest to the sun.
C. Milankovitch hypothesized that it is the amount of insolation reaching the polar regions in summer, and not minor differences in total global insolation, that induce large-scale climatic changes such as the ice ages.
D. Observations that periods of ice sheet advance and retreat correspond to various periods of the earth's orbital variation, along with current theoretical models of the earth's climate, support Milankovitch's hypothesis.
E. The increased content of CO_2 and other gases such as CFCs in the atmosphere have enhanced the greenhouse effect and could lead to global climate modification.

13-13. Ocean Basins

A. A **continental shelf** is a gently sloping surface extending seaward from a continental margin to a line marked by a sudden increase in slope.
B. A **continental slope** is a submerged, steeply sloping surface extending from the seaward edge of the continental shelf to the ocean floor.
C. The **continental rise** is the ocean floor beyond the base of the continental slope having a less pronounced slope.
D. The **abyssal plain** is the nearly level area of the ocean floor extending from the seaward margin of the continental rise.
E. The ocean basins have an average depth of 3.7 km.
F. The deepest point in the oceans occurs within the Marianas Trench and is 11 km below sea level.
G. The ocean floor consists of plains, valleys, isolated volcanic peaks, and mountain ranges such as the Mid-Atlantic Ridge.
H. The Antarctic ice cap contains about 90 percent of the world's permanent ice.
I. The world's oceans probably appeared about 4 billion years ago and have always been salty.

13-14. Ocean Currents

A. Oceans affect climate in two ways:
   1. Oceans act as heat reservoirs that moderate the seasonal temperature changes of adjacent land areas.
   2. Wind-driven ocean currents, by retaining the temperatures of their latitudes of origin, influence the temperatures of neighboring land areas.

B. Wind-driven surface currents parallel to a large extent the earth's major wind systems and aid the winds in their distribution of heat and cold over the surface of the earth.

C. Major ocean currents include the **equatorial current**, the **Gulf Stream**, the **Labrador Current**, and the **Japan Current**.

KEY TERMS AND CONCEPTS

The questions in this section will help you review the key terms and concepts from Chapter 13.

Multiple Choice

1. The layer of the atmosphere in which weather occurs is the
   a. mesosphere
   b. stratosphere
   c. thermosphere
   d. troposphere

2. The study of weather and weather patterns is called
   a. atmospherics
   b. climatology
   c. meteorology
   d. hydrology

3. Chlorofluorocarbons (CFCs) are considered to be among the worst pollutants because
   a. they reduce the total insolation reaching the earth's surface
   b. CFCs break down ozone molecules and weaken the ozone shield
   c. CFCs are partly responsible for acid rain.
   d. CFCs destroy the ionosphere and thus interfere with long-distance radio communication

4. Clouds are sometimes seeded to
   a. cause them to dissipate
   b. induce them to produce rain
   c. cause them to rise higher into the atmosphere
   d. cause them to produce lightning over uninhabited areas

5. The noon sun is lowest in the sky in the northern hemisphere and the period of daylight is shortest on about
   a. December 22
   b. March 21
   c. June 22
   d. September 23

6. Winds form in response to the
   a. uneven heating of the earth's surface
   b. rotation of the earth on its axis
   c. coriolis effect
   d. friction between ocean currents and the atmosphere

7. Seasonal large-scale sea and land breezes associated with pronounced wet and dry seasons are called
   a. doldrums
   b. westerlies
   c. trade winds
   d. monsoons

8. Which one of the following cloud types is normally associated with a cold front?
   a. nimbostratus
   b. cirrostratus
   c. altostratus
   d. cumulonimbus

9. Cold fronts typically
   a. move faster than warm fronts
   b. are less steep than warm fronts
   c. produce little or no precipitation
   d. have less of a temperature difference between opposing air masses

10. According to the Milankovitch theory of climatic change
    a. the sun's 11-year sunspot cycle is responsible for most climatic change
    b. the increase in atmospheric CO_2 has enhanced the greenhouse effect, resulting in global warming
    c. changes in sea level are responsible for climatic change
    d. variations in the amount of insolation reaching the polar regions in summer result in climatic change

11. Of the following divisions of the ocean floor, which is directly adjacent to a continental land mass?
   a. continental shelf
   b. continental rise
   c. continental slope
   d. abyssal plain

12. The ocean basins average ___________ km in depth.
   a. 11
   b. 3.7
   c. 0.8
   d. 2.4

True or False

Decide whether each statement is true or false. If false, either briefly state why it is false or correct the statement to make it true. See Chapter 1 or 2 for an example.

__________ 1. The two most abundant gases in the atmosphere are nitrogen and oxygen.

__________ 2. The amount of moisture that air can hold decreases with temperature.

__________ 3. Almost 80 percent of the insolation arriving at the upper atmosphere reaches the earth's surface.

__________ 4. The major source of atmospheric energy is long-wavelength infrared radiation from the earth, not direct sunlight.

__________ 5. Because the oceans cover most of the earth's surface, the ocean currents transport most of the earth's energy from one region to another.

__________ 6. Because of the coriolis effect, winds are curved to the left in the southern hemisphere.

__________ 7. A cyclone is an air mass in which the pressure is low at the center.

__________ 8. The earth's seasons are caused by shape of the earth's orbit.

__________ 9. Since the end of the last ice age, the earth's climate has remained remarkably stable.

__________ 10. The oceans act as heat reservoirs that moderate the climates of adjacent land areas.

Fill in the Blank

1. The ____________________ is the envelope of air surrounding the earth.

2. The ____________________ is the cloudless layer of the atmosphere directly above the troposphere.

3. The substance ____________________ in the earth's stratosphere absorbs harmful ultraviolet radiation.

4. A temperature ____________________ occurs when a layer of hot air aloft is warmer that the air immediately below.

5. Clouds are sometimes ____________________ with various substances such as silver iodide to induce rainfall.

6. Air is said to be ____________________ when it holds the maximum amount of water vapor at a given temperature.

7. ____________________ (2 words) is a percentage that indicates the extent to which air is saturated with water vapor.

8. ____________________ is the study of weather.

9. The ____________________ (2 words) results from the indirect heating of the atmosphere due to radiation of energy from the warm earth back into the atmosphere.

10. An ____________________ is a line of constant pressure on a weather map.

11. The ____________________ effect is the deflection of moving things such as air currents due to the earth's rotation.

12. An ____________________ (2 words) is a large body of air that is relatively uniform in temperature and moisture content.

13. The continental ____________________ extends seaward beyond the continental slope.

14. The ____________________ plains are the deepest portions of the ocean basins.

15. The ____________________ (2 words) were periods of severe cold in which ice sheets covered much of North America and Europe.

Matching

1._____ anticyclone

2._____ mesosphere

3._____ climate

4._____ cyclone

5._____ troposphere

6._____ weather

7._____ insolation

8._____ humidity

9._____ hydrosphere

10.____ doldrums

a. state of the atmosphere at a given region and time
b. includes all the water of the earth's surface
c. portion of the atmosphere between 50 and 80 km
d. high-pressure weather system
e. the water vapor content of the air
f. lowest region of the atmosphere
g. average weather conditions in an area over a period of years
h. solar energy that arrives at the earth's upper atmosphere
i. low-pressure weather system
j. region of light, erratic wind along the equator

WEB LINKS

The Climate Change Research Center provides information about climate change at

http://www.ccrc.sr.unh.edu/

Take a virtual tour of the world's oceans at the Ocean Planet exhibit at

http://seawifs.gsfc.nasa.gov/ocean_planet.html

Explore numerous ocean-related topics at the Scripps Institution of Oceanography at

http://www.sio.ucsd.edu/

Gain an understanding about weather phenomena and learn about severe weather at the National Severe Storms Laboratory at

http://www.nssl.noaa.gov/

ANSWER KEY

Multiple Choice

1. d 2. b 3. b 4. b 5. a 6. a 7. d 8. d 9. a 10. d 11. a 12. b

True or False

1. True
2. False. The amount of moisture that air can hold increases with temperature.
3. False. Slightly over half of the total insolation reaches the earth's surface. About 30 percent is reflected back into the space, and about 19 percent is absorbed by the atmosphere.
4. True
5. False. About 80 percent of the energy transported around the earth is carried by the winds. The remainder is carried by the ocean currents.
6. True
7. True
8. False. The earth's seasons are caused by the tilt of the earth's axis together with its annual orbit around the sun.
9. False. There have been several periods of rising and falling global temperatures since the end of the last ice age. We are now experiencing a period of worldwide warming.
10. True

Fill in the Blank

1. atmosphere
2. stratosphere
3. ozone
4. inversion
5. seeded
6. saturated
7. relative humidity
8. meteorology
9. greenhouse effect
10. isobar
11. coriolis
12. air mass
13. rise
14. abyssal
12. ice ages

Matching

1. d 2. c 3. g 4. i 5. f 6. a 7. h 8. e 9. b 10. j

Chapter 14

THE ROCK CYCLE

OUTLINE

GOALS

1. List in order of abundance the four chief elements in the earth's crust.
2. Explain why the silicates can vary so much in composition and crystal structure.
3. Distinguish between rocks and minerals.
4. Briefly describe quartz, feldspar, mica, the ferromagnesian minerals, the clay minerals, and calcite.
5. Distinguish among igneous, sedimentary, and metamorphic rocks.
6. Compare the origins of the fine-grained and coarse-grained igneous rocks and give several examples of each type.
7. Describe several fragmental sedimentary rocks.
8. State the main characteristics of limestone. How is it formed?
9. Describe several metamorphic rocks and give their origins.
10. Distinguish among the four kinds of earthquake waves.
11. Explain the evidence that suggests the division of the earth into core, mantle, and crust.
12. Give several reasons for the belief that earth's core is largely molten iron.
13. Identify the main source of heat that flows out of the earth's interior.
14. Compare the earth's magnetic field with the magnetic field of a bar magnet, and explain why no actual permanent magnet can give rise to the earth's field.
15. Describe the chemical and mechanical weathering of rocks.
16. Discuss the development of a valley carved by a river.
17. Discuss the development of a valley carved by a glacier.
18. Define groundwater, saturated zone, water table, spring, and aquifer.
19. Discuss the deposition of stream and glacier sediments.
20. Describe the processes by which sediments become rock.
21. Describe the events that occur in a typical volcanic eruption.

22. Indicate on a map of the world the regions where volcanoes and earthquakes occur most often.
23. Give the reasons why granite is believed to have hardened from a molten state underground.
24. Draw a diagram that shows the rock cycle.

CHAPTER SUMMARY

The structure of our planet and the materials that compose the earth's **crust** are the main topics of Chapter 14. The term **mineral** is defined, and six common minerals are described in terms of their chemical compositions, physical properties, and uses. The distinction between **igneous**, **metamorphic**, and **sedimentary** rocks is made, and examples of rocks from each category are discussed in terms of their formation and physical characteristics. Our knowledge of the earth's internal structure is shown to be based on analysis of the behavior of earthquake waves. What drives the earth's internal heat engine and the origin of the earth's magnetic field are revealed.
The processes of **erosion**, **weathering**, **sedimentation**, **volcanism**, and **diastrophism** are related to the rock cycle and the changing nature of the earth's crust.

CHAPTER OUTLINE

14-1. Composition of the Crust

A. The outer part of the earth is called the **crust**.
B. The two most abundant elements of the earth's crust are oxygen and silicon.
C. Most crustal rocks are composed of silicon compounds.
   1. **Silica** is silicon dioxide (SiO_2). **Silicates** are compounds of silicon with oxygen and one or more metals.
   2. Silicates are crystalline solids whose basic structural unit is the SiO_4^{4-} tetrahedron.
   3. The wide variety of silicate minerals is due to many different ways the SiO_4^{4-} tetrahedron can combine with metal ions.

14-2. Minerals

A. **Minerals** are naturally occurring, crystalline, inorganic solids having fairly specific chemical compositions.
B. More than 2000 minerals are known, but most are rare.
C. Some physical properties of minerals are color, hardness, **crystal form**, and **cleavage** (tendency to split along certain planes).
D. Examples of six common minerals are:
   1. **Quartz** (SiO_2) crystals are often found in narrow deposits called **veins**.
   2. **Feldspars** are silicate minerals having very similar properties and are the most abundant single constituent of rocks. Varieties include orthoclase (silicate of K and Al) and plagioclase (silicate of Na, Ca, and Al).

3. **Mica** varieties include white mica (silicate of H, K, and Al) and black mica (silicate of H, K, Al, Mg, and Fe).
4. **Ferromagnesian minerals** are silicates of Fe, Mg, and usually other metals such as K, Al, and Ca.
5. **Clay minerals** are silicates of Al, some with a little Mg, Fe, and K.
6. **Calcite** ($CaCO_3$) is the chief mineral in limestone and marble.

14-3. Igneous Rocks

A. **Igneous rocks**, such as granite, basalt, and **obsidian** (natural glass), are rocks that have cooled from a molten state.

B. Fine-grained igneous rocks (examples: rhyolite, andesite, basalt) have cooled rapidly at or near the earth's surface. Coarse-grained ones (examples: granite, diorite, gabbro) have cooled slowly well below the earth's surface.

14-4. Sedimentary Rocks

A. **Sedimentary rocks** have consolidated from materials derived from the disintegration or solution of other rocks and deposited by water, wind, or glaciers.

B. Sedimentary rocks are divided into two categories:
   1. Fragmental rocks
   2. Chemical and biochemical precipitates

C. Three types of fragmental sedimentary rocks are:
   1. **Conglomerate** (cemented gravel)
   2. **Sandstone** (cemented sand grains)
   3. **Shale** (consolidated mud or silt)

D. Examples of precipitated sedimentary rocks are:
   1. **Limestone** (mostly the mineral calcite)
   2. **Chalk** (loosely consolidated variety of limestone)
   3. **Chert** (**flint** and **jasper** are varieties)

14-5. Metamorphic Rocks

A. **Metamorphic rocks** are formed from preexisting rocks that have been altered by heat and/or pressure.

B. Many metamorphic rocks display **foliation**, which is an arrangement of flat or elongated mineral grains in parallel layers that give the rocks a banded or layered appearance.

C. Examples of foliated metamorphic rocks are:
   1. **Slate** (weakly metamorphosed shale)
   2. **Schist** (severely metamorphosed shale or fine-grained igneous rock)
   3. **Gneiss** (severely metamorphosed rocks, except pure limestone and pure quartz sandstone)

D. Examples of nonfoliated or weakly foliated metamorphic rocks are:
    1. **Marble** (metamorphosed limestone)
    2. **Quartzite** (metamorphosed sandstone)

14-6. Earthquakes

A. Most **earthquakes** are due to the sudden movement of rock along fracture surfaces called **faults**.
B. Earthquake magnitudes are expressed on the Richter scale.
C. Reliable earthquake predictions do not exist.

14-7. Structure of the Earth

A. The two main categories of earthquake waves are "body waves" (which move through the earth's interior) and "surface waves" (which travel along the earth's surface).
B. Types of body waves include:
    1. Primary (P) waves, which are longitudinal or pressure waves and can travel through a liquid.
    2. Secondary (S) waves, which are transverse waves and cannot travel through a liquid.
C. Types of surface waves include:
    1. Love waves, which are transverse waves in which the earth's surface vibrates from side to side.
    2. Rayleigh waves, which involve orbital motions that cause the earth's surface to move up and down..
D. P waves travel faster than S waves, and surface waves are the slowest.
E. The earth is composed of a series of layers:
    1. The **crust** (the outer layer)
    2. The **mantle** (the layer beneath the crust)
    3. The **core** (the earth's innermost zone)
F. The internal structure of the earth has been determined with the help of P and S earthquake waves.
    1. P and S waves are refracted as they travel within the earth.
    2. P and S waves do not reach **shadow zones** on the other side of the earth from their origins. This is evidence for a central core.
    3. S waves cannot travel through the core, suggesting that at least the outer core is liquid.
    4. Faint traces of P waves in the shadow zones suggest that the earth's inner core is solid.
G. The lower boundary of the crust is known as the **Mohorovicic discontinuity**.

14-8. The Earth's Interior

A. The upper mantle appears to be composed mainly of ferromagnesian minerals.
B. Enormous pressures in the lower mantle have compressed minerals into crystal forms that are the most compact possible including very dense oxides of Si, Fe, and Mg.
C. The liquid outer core and the solid inner core are believed to be composed mainly of iron and nickel.
D. Temperatures inside the earth are believed to range from 375°C at the top of the mantle to at least 3000°C at the core's center and possibly much more.
E. Most of earth's interior heat is due to radioactivity.

14-9. Geomagnetism

A. In 1600, Sir William Gilbert proposed that earth behaves like a giant magnet.
B. The earth's magnetic field is believed to arise from coupled fluid motions and electric currents in the liquid iron of the core.

14-10. Weathering

A. **Erosion** includes all the processes by which rocks are worn down and the debris carried away.
B. **Weathering** is the gradual disintegration of exposed rocks.
   1. **Chemical weathering** refers to the chemical decomposition of rock.
   2. **Mechanical weathering** refers to the physical disintegration of rock.
C. Soil formation is an important result of weathering.

14-11. Stream Erosion

A. Running water is the main agent of erosion.
B. In a young landscape, a stream carves a narrow V-shaped valley.
C. With time, continued stream erosion results in a **flood plain** with a flat floor.
D. In a mature landscape, a treelike drainage pattern of secondary streams develops.
E. In a old landscape, flood plains are broad and further erosion is slow.
F. Stream erosion would ultimately reduce the land surface to a flat plain if no other processes such as geologic uplift took place.

14-12. Glaciers

A. A moving mass of ice formed by accumulated snow is called a **glacier**.
   1. **Valley glaciers** are glaciers lying in mountain valleys. Valleys carved by valley glaciers have U-shaped cross sections.

2. **Continental glaciers** or **ice caps** are domal masses of ice that cover large areas of the earth's surface.

B. Rock fragments imbedded in the bottom of a glacier cut and scrape the underlying rock, resulting in glacial erosion.

14-13. Groundwater

A. All the water that penetrates the earth's surface is called **groundwater**.

B. When spaces within the soil and any underlying porous rocks are filled with water, the ground is said to be **saturated**.

C. The **water table** is the upper surface of the saturated zone .

D. A **spring** is formed where groundwater from the saturated zone comes to the surface.

E. An **aquifer** is a body of porous rock through which groundwater moves.

14-14. Sedimentation

A. The eroded material transported by the agents of erosion is eventually deposited to form **sediments**.

B. The most widespread sediments collect near continental margins.

C. There are four common sites of deposition:

1. Flood-deposited debris in stream gravel banks and sandbars
2. The flood plains of meandering rivers
3. **Alluvial fans**, which are deposits of sediments where streams emerge from steep mountain valleys and flow onto plains
4. **Deltas,** which are deposits of sediments where a stream enters a lake or sea

D. **Moraines** are piles of debris that accumulate around the ends and along the sides of glaciers. This material is called **till**.

E. The most important agents of deposition are ocean currents because of the large volume of sediment they carry and deposit.

F. Examples of groundwater deposition include:

1. Deposition of minerals in veins within rock
2. Cave deposits including **stalactites** (hanging from cave ceilings) and **stalagmites** (rising from cave floors)
3. Deposits around hot springs and geysers

G. **Lithification** is the process by which sediments become rock.

H. Lithification includes:

1. Compaction, in which the sediment grains are squeezed together under the pressure of overlying deposits
2. Concentration, in which the sediment grains are bound together by chemical changes brought about by circulating groundwater

14-15. Volcanoes

A. The processes of **vulcanism** (the movements of molten rock) and **diastrophism** or **tectonism** (the movements of the solid materials of the earth's crust) act in opposition to the processes that would level the earth's surface.
B. A **volcano** is an opening in the earth's crust through which molten rock (called **magma** while underground, **lava** above ground) pours forth.
C. A volcano usually has a depression, or a **crater**, at its summit.
D. The two main factors that determine whether an eruption will be quiet or explosive are:
    1. **Viscosity** (resistance to flow) of the magma
    2. Amount of dissolved gases, such as water vapor (the most prominent), carbon dioxide, nitrogen, hydrogen, and various sulfur compounds in the magma
E. Magmas rich in silica (the most viscous) and dissolved gases result in explosive eruptions, while magmas with modest gas and silica contents result in quiet eruptions.
F. Lava hardens into one or another type of volcanic rock including:
    1. Basalt (the most common)
    2. Rhyolite (the most silica rich)
    3. **Pumice** (light and porous)
G. Most active volcanoes occur around the borders of the Pacific Ocean, on some of the Pacific Islands, in Iceland, and in East Africa.

14-16. Intrusive Rocks

A. **Plutons** are intrusive bodies formed by the solidification of magma under the earth's surface.
B. Because plutons cool slowly, the resulting intrusive igneous rocks tend to be coarse-grained. An example is granite.
C. A **dike** is a wall of intrusive igneous rock that cuts across existing rock layers.
D. A **sill** is a pluton formation that lies between and parallel to existing rock strata.
E. A **laccolith** is a pluton formation that forms a mushroom-shaped intrusion that pushes up overlying rock strata.
F. A **batholith** is a very large pluton that can cover hundreds of thousands of square km. Batholiths are always associated with mountain ranges, past or present.

14-17. The Rock Cycle

A. Rocks can change from one kind to another in a variety of ways.
B. The rock cycle is a never-ending process.

KEY TERMS AND CONCEPTS

The questions in this section will help you review the key terms and concepts from Chapter 14.

Multiple Choice

Circle the best answer for each of the following questions.

1. In the silicates, the basic silicon-oxygen structural element is in the shape of a
   a. cube
   b. rhombohedron
   c. octahedron
   d. tetrahedron

2. The most abundant single constituent of rocks is
   a. quartz
   b. calcite
   c. white mica
   d. feldspar

3. A group of fibrous silicate minerals once widely used because of their mechanical strength and resistence to fire but now known to cause lung and intestinal cancers is
   a. feldspar
   b. asbestos
   c. cinnabar
   d. satin spar

4 A mineral used as an insulator in electrical equipment is
   a. kaolin
   b. white mica
   c. calcite
   d. quartz

5. Limestone and marble are composed chiefly of the mineral
   a. calcite
   b. mica
   c. quartz
   d. olivine

6. A shiny, black igneous rock having the texture of glass is
   a. slate
   b. conglomerate
   c. obsidian
   d. schist

7. A coarse-grained metamorphic rock that resembles granite except for its banded appearance is
   a. chert
   b. quartzite
   c. slate
   d. gneiss

8. A sedimentary rock formed either as a chemical precipitate or by the consolidation of shell fragments is
   a. limestone
   b. gneiss
   c. granite
   d. sandstone

9. A seismograph detects
   a. temperature gradients within the earth
   b. earthquake waves
   c. fluctuations in the earth's magnetic field
   d. radioactivity

10. S waves differ from P waves in that
   a. S waves can travel through liquids and P waves cannot
   b. S waves are body waves and P waves are surface waves
   c. S waves travel faster than P waves
   d. S waves are transverse waves and P waves are longitudinal waves

11. The earth's core
   a. consists of a solid outer core and a liquid inner core
   b. is composed mainly of silicon
   c. is the source of the earth's magnetic field
   d. is completely solid

12. Till is
   a. a type of volcanic rock
   b. all of the material deposited directly by a glacier
   c. material deposited by a river within its floodplain
   d. a mineral deposit associated with hot springs and geysers

13. A body of porous rock through which groundwater moves is called an
    a. aquitard
    b. aquifer
    c. aquiclude
    d. aqueduct

14. The process by which sediments become rock is called
    a. diastrophism
    b. compaction
    c. tectonism
    d. lithification

15. Which one of the following statements about batholiths is false?
    a. The principle rock in batholiths is granite.
    b. Batholiths are always associated with mountain ranges, either past or present.
    c. Batholiths are very large plutons.
    d. Batholiths form when magma cools rapidly at or very near the earth's surface.

True or False

Decide whether each statement is true or false. If false, either briefly state why it is false or correct the statement to make it true. See Chapter 1 or 2 for an example.

__________ 1. The three most abundant elements by mass within the earth's crust are oxygen, silicon, and carbon.

__________ 2. Coarse-grained igneous rocks such as granite indicate a rapid cooling from a molten state.

__________ 3. Fossils are most frequently found in sedimentary rock.

__________ 4. The shadow zone separates P waves that have reached the surface through the mantle from P waves that have reached the surface after passing through both the mantle and the core.

__________ 5. The earth's core makes up about 55 percent of the earth's volume.

__________ 6. Changes in the patterns of flow of the liquid iron in the earth's core are believed to have caused reversals in the earth's magnetic field.

__________ 7. In a mature landscape, valleys are few and have characteristic V-shaped channels.

__________ 8. An alluvial fan forms where a stream's flow is stopped abruptly as the stream enters a lake or the sea.

__________ 9. Magmas rich in silica and dissolved gases are associated with quiet eruptions.

__________ 10. Pumice is the most common volcanic rock.

Fill in the Blank

1. The ____________________ of the earth is its outer shell of rock.

2. ____________________ are compounds of silicon with oxygen and one or more metals.

3. A ____________________ is a naturally occurring crystalline inorganic solid that has a fairly specific chemical composition.

4. ____________________ is the tendency of a mineral to split along certain planes.

5. The ____________________ of an earthquake is the point on the earth's surface directly over the quake's focus.

6. The ____________________ discontinuity is the crust's lower boundary.

7. ____________________ weathering involves the physical disintegration of rock.

8. A ____________________ is the wide, flat floor of a river valley.

9. ____________________ are moving masses of ice formed from accumulated snow.

10. ____________________ is water that penetrates earth's surface and soaks into the ground.

11. ____________________ are cave deposits that hang from the roofs of limestone caves; ____________________ are cave deposits that rise from cave floors.

12. ____________________, or tectonism, involves the movements of the solid materials of the earth's crust.

13. Molten rock is usually called ____________________ while underground and is called ____________________ aboveground.

14. A ____________________ is an intrusive body of igneous rock that formed from molten rock that rose through the earth's crust but solidified before reaching the surface.

15. A ____________________ is a wall of igneous rock that cuts across existing rock layers.

Matching

1._____	quartz	a.	metamorphosed sandstone
2._____	marble	b.	consolidated variety of limestone
3._____	schist	c.	porous volcanic rock than can float on water
4._____	pumice	d.	coarse-grained igneous rock
5._____	quartzite	e.	metamorphosed limestone
6._____	feldspars	f.	most abundant single constituent of rocks
7._____	conglomerate	g.	coarse sedimentary rock consisting of cemented gravel
8._____	chalk	h.	foliated metamorphic rock
9._____	shale	i.	consolidated mud or silt
10.____	granite	j.	mineral composed of silicon dioxide

WEB LINKS

Take a virtual tour of the San Andreas fault at

http://sepwww.stanford.edu/oldsep/joe/fault_images/BayAreaSanAndreasFault.html

Explore the Tech Museum of Innovation Earthquake On-line Exhibit at

http://www.thetech.org/exhibits_events/online/quakes

Take a virtual field trip of Hawaii's Kilauea volcano at

http://satftp.soest.hawaii.edu/space/hawaii/vfts/kilauea/kilauea.vfts.html

Volcano World is a comprehensive source of information about volcanoes. Visit the site at

http://volcano.und.nodak.edu/vw.html

ANSWER KEY

Multiple Choice

1. d 2. d 3. b 4. b 5. a 6. c 7. d 8. a 9. b 10. d 11. c 12. b 13. b 14. d 15. d

True or False

1. False. The three most abundant elements by mass within the earth's crust are oxygen, silicon, and aluminum.
2. False. Coarse-grained igneous rocks such as granite indicate a slow cooling from a molten state.
3. True
4. True
5. False. The earth's core makes up less than 20 percent of its volume.
6. True
7. False. In a young landscape, valleys are few and have characteristic V-shaped channels.
8. False. A delta forms where a stream's flow is stopped abruptly as the stream enters a lake or the sea.
9. False. Magmas rich in silica and dissolved gases are associated with explosive eruptions.
10. False. Basalt is the most common volcanic rock.

Fill in the Blank

1. crust
2. silicates
3. mineral
4. cleavage
5. epicenter
6. Mohorovicic
7. mechanical
8. floodplain
9. glaciers
10. groundwater
11. stalactites, stalagmites
12. diastrophism
13. magma, lava
14. pluton
15. dike

Matching

1. j 2. e 3. h 4. c 5. a 6. f 7. g 8. b 9. i 10. d

Chapter 15

THE EVOLVING EARTH

OUTLINE

Tectonic Movement
15.1 Types of Movement
15.2 Mountain Building
15.3 Continental Drift

Plate Tectonics
15.4 Lithosphere and Asthenosphere
15.5 The Ocean Floors
15.6 Ocean-Floor Spreading
15.7 Plate Tectonics

Methods of Historical Geology
15.8 Principles of Uniform Change
15.9 Rock Formations
15.10 Radiometric Dating
15.11 Fossils
15.12 Geologic Time

Earth History
15.13 Precambrian Time
15.14 The Paleozoic Era
15.15 Coal and Petroleum
15.16 The Mesozoic Era
15.17 The Cenozoic Era
15.18 Human History

GOALS

1. Describe the various kinds of tectonic movement.
2. List the main steps in the development of the great mountain ranges of the world.
3. Outline the evolution of today's continents from the former supercontinent Pangaea.
4. Distinguish between the lithosphere and the asthenosphere.
5. Explain how ocean-floor spreading accounts for present-day features of the ocean floors.
6. Describe what happens at oceanic-continental, oceanic-oceanic, and continental-continental plate collisions.
7. Identify the San Andreas Fault and indicate its origin.
8. State what is meant by the principle of uniform change.
9. List the four basic principles of historical geology.
10. Describe how radiometric dating is used to find the ages of rocks.
11. Describe how radiocarbon dating is used to find the ages of biological specimens.
12. Explain why animal fossils are more common than plant fossils, and account for the abundance of fossils on the floors of ancient shallow seas.
13. Discuss the various ways in which fossils are useful in geology.
14. Give the basis for the division of geological time into eras, periods, and epochs.
15. List the four major divisions of geological time in order from past to present.
16. Describe the formation of coal and petroleum.
17. Identify dinosaurs and discuss what may have led to their sudden disappearance.
18. Account for the survival of mammals when the dinosaurs disappeared.
19. Discuss the most recent Ice Age and its role in early human history.
20. Explain why overpopulation is the principle hazard facing the world of the future.

CHAPTER SUMMARY

We live on a dynamic, continually changing planet earth. Why this is so and how today's earth differs from the earth of the distant past are the main topics of Chapter 15. Movements of the earth's crust, including **faulting**, **folding**, and **mountain building,** are explained in terms of the theory of **plate tectonics**. The tools geologists use to reconstruct past geologic events and an overview of earth's history are presented. The origins of coal and petroleum are revealed, and theories concerning the extinction of the dinosaurs are discussed. The extent and impacts of the Ice Age in North America are presented. The environmental problems associated with global overpopulation are emphasized.

CHAPTER OUTLINE

15-1. Types of Deformation

A. A **fault** is a rock fracture along which one side has moved relative to the other.
B. **Fault scarps** are cliffs formed by faulting.
C. **Folding** occurs when rock strata are compressed by slow, continuous tectonic movements of the crust.
D. Large-scale crustal movements (called **tectonic** changes) involve the rising, falling, tilting, or warping of whole continents or large parts of them.

15-2. Mountain Building

A. Mountains form in a variety of ways.
   1. Mountains can form due to volcanic activity.
   2. Some mountains are blocks of the earth's crust raised along faults.
   3. The great mountain ranges have a long, complex history involving sedimentation, folding, faulting, igneous activity, repeated uplifts, and deep erosions.
B. In mountains having a history of sedimentation:
   1. The sedimentary layers are usually thicker than similarly aged sedimentary layers under adjacent plains.
   2. The sedimentary rocks have a complex structure resulting from intensive compressional forces.

15-3. Continental Drift

A. The German meteorologist Alfred Wegener proposed the theory of continental drift early in this century.
B. The evidence Wegener cited in support of continental drift included:
   1. The occurrence of similar fossils on widely separated landmasses
   2. Data on prehistoric climates that indicated that continents had changed position in latitude with time

C. Wegener suggested that the continents were originally part of a single, huge landmass called **Pangaea**.

D. About 200 million years ago, Pangaea broke apart into two supercontinents called **Laurasia** and **Gondwana**.

  1. Laurasia comprised what is now North America, Greenland, and most of Eurasia.
  2. Gondwana comprised what is now South America, Africa, Antarctica, India, and Australia.

E. Laurasia and Gondwana were separated by the **Tethys Sea**.

F. Both Laurasia and Gondwana broke up to give rise to today's continents.

15-4. Lithosphere and Asthenosphere

A. The **lithosphere** is a shell of hard rock that consists of the crust and the upper mantle.

B. The **asthenosphere** is a layer of hot, soft rock lying just below the lithosphere.

15-5. The Ocean Floors

A. Methods used to study the ocean floors include:

  1. Echo sounding
  2. Core sampling
  3. Determination of magnetic properties of ocean floor rocks

B. Four findings about ocean floors have clarified our understanding about crustal evolution.

  1. The ocean floors are geologically very young (200 million years or less in age).
  2. A worldwide system of ocean **ridges** and **rises** runs across the ocean floors.
  3. A system of deep **trenches** rims much of the Pacific Ocean. Some trenches have volcanic **island arcs** on their landward sides.
  4. Ocean-floor rocks are magnetized in the same direction in strips parallel to the midocean ridges; however, the direction is reversed from strip to strip going away from a ridge on either side.

15-6. Ocean-Floor Spreading

A. In the early 1960s, the American geologists Harry H. Hess and Robert S. Deitz independently proposed that the ocean floors are spreading.

B. During the process of ocean-floor spreading, molten rock rises up along a midocean ridge and pushes apart the lithosphere on either side of the ridge. The molten rock hardens and becomes new crust.

C. The discovery of normal and reverse magnetic strips on either side of an ocean ridge is strong evidence for ocean-floor spreading.

15-7. Plate Tectonics

A. According to the theory of **plate tectonics**, the lithosphere is divided into seven huge **plates** and a number of smaller ones.
B. New lithosphere is created where plates move apart at midocean ridges.
C. There are three kinds of plate collisions:
    1. **Ocean-continental plate collision** occurs where a plate bearing dense oceanic crust slides underneath a plate bearing lighter continental crust in a region of contact called a **subduction zone.**
    2. **Oceanic-oceanic plate collision** occurs where a plate bearing oceanic crust slides underneath a second plate bearing oceanic crust in a subduction zone.
    3. **Continental-continental plate collision** occurs when the plate edges are pushed together and buckle, forming a mountain range.
D. A **transform fault** is a type of plate boundary where the edges of two plates slide past each other.
    1. The San Andreas Fault is an example of a transform fault that has formed between the boundary of the Pacific and North American plates.
    2. The San Francisco earthquakes of 1906 and 1989 were caused by movement along the San Andreas Fault.
E. Two plausible mechanisms for plate motion have been proposed.
    1. Convection in the asthenosphere
    2. Subduction of oceanic plate margins
F. Thirty million years from now, if present plate motions continue:
    1. The Atlantic Ocean will widen.
    2. The Pacific Ocean will narrow.
    3. Part of California will detach from North America.
    4. The Arabian peninsula will become part of Asia.
    5. The West Indies will become a land bridge between the Americas.
    6. The western Pacific islands will increase in extent.
G. The earth's surface will likely continue to change in the future.

15-8. Principle of Uniform Change

A. Seventeenth-century theologian Bishop Ussher, using stories in the Bible, determined that earth was created at 9 o'clock in the morning of October 12, 4004 B.C.
B. The German geologist Abraham Gottlob Werner believed that all rocks were sedimentary rocks and that the geologic history of the earth consisted of three sudden precipitations from an ancient ocean that were followed by the disappearance of the water.
C. The French biologist Georges Cuvier regarded the earth's history as a succession of catastrophes. Cuvier based his ideas of earth's history on the study of **fossils,** the remains or traces of organisms preserved in rocks.
D. The Scottish geologist James Hutton proposed that earth's history could be understood in terms of processes under way in the present-day world.

E. The English geologist Charles Lyell (1797–1875) modified and expanded Hutton's ideas and proposed the **principle of uniform change**, which stated that geologic processes in the past were the same as those in the present.

F. Charles Darwin's **theory of evolution** showed that changes in living things as well those in the inorganic world of rocks could be explained in terms of processes operating all around us.

15-9. Rock Formations

A. It is often possible to reconstruct past geologic events in terms of processes still at work reshaping the earth's surface.

B. The science of geology is faced with two fundamental problems:
   1. To arrange in order the events recorded in the rocks of a single outcrop or small region
   2. To correlate events in various regions of the world to give a connected history of the earth

C. Basic principles of historic geology include:
   1. In a sequence of sedimentary rocks, the lowest bed is the oldest and the highest bed is the youngest.
   2. Sedimentary beds were originally deposited in more or less horizontal layers.
   3. Tectonic movement took place after the deposition of the youngest bed affected.
   4. An igneous rock is younger than the youngest bed it intrudes.

D. An **unconformity** is a buried surface of erosion and involves at least four geologic events:
   1. The deposition of the oldest strata
   2. Tectonic movement that raises and perhaps tilts the existing strata
   3. Erosion of the elevated strata to produce an irregular surface
   4. A new period of deposition that buries the eroded surface

15-10. Radiometric Dating

A. Two methods are used to determine the worldwide sequence of events that have shaped the earth's surface:
   1. Radioactive dating
   2. Fossil identification

B. The decay of a particular radionuclide proceeds at a steady rate, and the ratio between the amounts of that nuclide and its stable daughter in a rock sample indicates the age of the rock.

C. The carbon isotope ^{14}C, called **radiocarbon**, makes it possible to evaluate the ages of ancient objects (wooden implements, cloth, leather, charcoal from campfires, etc.) and remains of organic origin.

15-11. Fossils

A. The most common fossils are the hard parts of animals (shells, bones, teeth, etc.). Plant fossils are relatively scarce.
B. Some fossils are trails or footprints left in soft mud and covered by later sediments.
C. Fossils are preserved under special conditions of burial.
D. Rock layers from different periods can be recognized by the kinds of fossils they contain, making possible the arrangement of strata in a relative time sequence.
E. Fossils help to reconstruct the environment in which the organisms lived.

15-12. Geologic Time

A. The past 570 million years of earth's history are divided into three time divisions called **eras**.
    1. **Cenozoic** Era (began 65 million years before present)
    2. **Mesozoic** Era (began 225 million years before present)
    3. **Paleozoic** Era (began 570 million years before present)
B. The nearly 4 billion years before the Paleozoic is usually considered as a single long division called **Precambrian time**.
C. The fossil record reveals a number of biological **extinctions** that have been used to divide geologic history into **periods**.
D. Periods are subdivided into shorter time spans called **epochs**.

15-13. Precambrian Time

A. Precambrian rocks often show considerable metamorphism and have been greatly deformed.
B. Geologic processes occurring during Precambrian time resemble those of today.
C. Precambrian rocks are exposed over a broad area covering most of eastern Canada and adjacent parts of the United States.
D. Indirect evidence suggests some type of life existed at least 3.7 billion years ago.
E. Primitive blue-green algae (cyanobacteria) were among the earliest organisms. Their photosynthetic activity provided the atmosphere with its oxygen.
F. Fossils of multicellular algae and fungi have been discovered in the 1.9-billion-year-old Gunflint Formation in Ontario, Canada.

15-14. The Paleozoic Era

A. Marine **invertebrates** (animals lacking internal skeletons) are the oldest fossils of the Paleozoic Era.
B. Fossils in late Paleozoic rocks show the first evidence of land-dwelling organisms, including primitive plants (tree ferns, land snails, primitive insects).

C. The first amphibians appeared in the Devonian Period; the first reptiles in the Carboniferous.
D. Reptiles became the dominant land creatures during the Permian Period.
E. The end of the Paleozoic Era was a time of intense tectonic activity. The Appalachian trough was compressed and uplifted into a mountain chain.

15-15. Coal and Petroleum

A. During the Carboniferous Period, much coal was formed in the present eastern United States and Europe.
B. Coal beds represent the sites of ancient Paleozoic swamps.
C. Coal was formed when accumulated plant matter (largely cellulose) underwent slow bacterial decay.
   1. Decay took place underwater and in the absence of air.
   2. Heat and pressure resulting from burial beneath later sediments converted the plant carbon into coal.
D. Petroleum originated from marine life such as plankton and algae.
E. Three steps are involved in the conversion of organic material into petroleum:
   1. Bacterial decay in the absence of oxygen
   2. Burial and modification by low-temperature chemical reactions
   3. "Cracking" of complex hydrocarbons to straight-chain alkane hydrocarbons (at higher temperatures natural gas forms instead of petroleum)
F. Petroleum and natural gas can migrate through porous rocks and become available for recovery when trapped by anticlines or beneath layers of impermeable clays or shales.

15-16. The Mesozoic Era

A. Pangaea split into Laurasia and Gondwana (followed by their own breakup) during the Mesozoic Era.
B. During the early Mesozoic Era, North America separated from Laurasia.
C. The **dinosaurs** were the dominant land animals during the Mesozoic Era.
D. Flowering plants first appeared in the mid-Mesozoic.
E. The first true birds arose from reptilian ancestors in the Jurassic Period.
F. **Mammals** first appeared in the Triassic Period.
G. The dinosaurs disappeared during a time of mass extinctions that divides the Mesozoic from the Cenozoic.
H. Theories for the mass extinctions include:
   1. Comet or meteorite impact
   2. Intense volcanic activity

15-17. The Cenozoic Era

A. Widespread volcanic activity and tectonic disturbance characterize Cenozoic time.
   1. The Alps, Carpathians, and Himalayas were uplifted in the mid-Tertiary Period.
   2. Late Tertiary saw the formation of the Cascade range.
   3. The Appalachians, Rockies, and Sierra Nevada were repeatedly uplifted and eroded, creating their present topography.

B. In the Cenozoic, the continents continued their earlier drifts, resulting in the present configuration of the continents.
C. After the extinction of the dinosaurs, mammals multiplied and evolved rapidly.
D. By the end of the Tertiary both the physical and organic worlds more closely resembled their present aspect.
C. In the Cenozoic, the biological isolation of the continents increased.
D. The Pleistocene Epoch saw the formation of ice caps in Canada and northern Europe.
E. In North America, the ice spread outward from three centers of accumulation in Canada.

15-18. Human History

A. Fossil evidence indicates that the human species had its origin in eastern Africa.
B Humans entered North America at least 15-20,000 years ago.
A. Human activities and overpopulation endanger the future.
B. Overpopulation is the main hazard facing the world.
C. Global population can be stabilized at a reasonable level, but action must be taken now while there is still time.

KEY TERMS AND CONCEPTS

The questions in this section will help you review the key terms and concepts from Chapter 15.

Multiple Choice

1. Cliffs formed as a result of faulting are called
   a. synclines
   b. fault scarps
   c. fault cliffs
   d. fault walls

2. The German meteorologist who proposed the theory of continental drift was
   a. Alfred Wegener
   b. William Herschel
   c. Friedrich Bessel
   d. Victor Hess

3. Chesapeake Bay is an example of a(n)
   a. delta
   b. water table
   c. drowned valley
   d. aquifer

4. The breakup of Pangaea began about
   a. 30 million years ago
   b. 60 million years ago
   c. 145 million years ago
   d. 200 million years ago

5. Present-day continents that were once part of Gondwana do not include
   a. South America
   b. Africa
   c. Australia
   d. Greenland

6. The crust and the outermost part of the mantle make up the
   a. asthenosphere
   b. outer core
   c. lithosphere
   d. Moho

7. Compared to continental crust, ocean floor crust is
   a. much younger
   b. much older
   c. slightly older
   d. about the same age

8. All the following geological features could be formed where plates collide except
   a. midocean ridges
   b. mountain ranges
   c. volcanoes
   d. island arcs

9. The type of fault that occurs where two plates slide past each other is a
   a. normal fault
   b. thrust fault
   c. transform fault
   d. reverse fault

10. At a subduction zone
    a. the sea floor spreads apart
    b. one lithospheric plate is forced under another
    c. molten rock rises to the surface
    d. new crust is forming

11. New lithosphere is created at _______ and destroyed at _______.
    a. midocean ridges, subduction zones
    b. subduction zones, midocean ridges
    c. transform faults, unconformities
    d. unconformities, transform faults

12. The Himalayas were formed as a result of
    a. diverging lithospheric plates
    b. oceanic-continental plate collision
    c. oceanic-oceanic plate collision
    d. continental-continental plate collision

13. Bishop James Ussher used biblical accounts to determine that the earth was created in the year
    a. 6002 B.C.
    b. 4004 B.C.
    c. 1604 B.C.
    d. 2006 B.C.

14. The ratio of ^{14}C to ^{12}C in a sample of dead animal or plant tissue indicates
    a. how long the animal or plant lived
    b. the elapsed time since the death of the organism
    c. how rapidly radiocarbon is expelled from the organism's tissues
    d. the tolerance of the organism to radioactive ^{14}C

15. The phrase "Cambrian explosion" refers to
    a. extensive volcanism at the end of the Cambrian period
    b. a mass extinction at the end of the Cambrian period triggered by a meteorite impact
    c. an increase in the diversity of living organisms beginning in the Cambrian period
    d. the rapid accumulation of molecular oxygen in the earth's atmosphere

16. Which one of the following represents the longest time span?
    a. Cenozoic Era
    b. Paleozoic Era
    c. Mesozoic Era
    d. Precambrian time

17. If you could travel back in time to the time of the dinosaurs, in which unit of geologic time would you be?
    a. Cenozoic Era
    b. Mesozoic Era
    c. Paleozoic Era
    d. Carboniferous Period

18. Pterosaurs were
    a. the ancestors of modern birds
    b. flying reptiles
    c. marine reptiles
    d. small, carnivorous dinosaurs

19. Pangaea split into Laurasia and Gondwana during
    a. Precambrian time
    b. the Paleozoic Era
    c. the Mesozoic Era
    d. the Cenozoic Era

20. The Ice Ages occurred during the
    a. Late Cretaceous Period
    b. Paleocene Epoch
    c. Pliocene Epoch
    d. Pleistocene Epoch

True or False

Decide whether each statement is true or false. If false, either briefly state why it is false or correct the statement to make it true. See Chapter 1 or 2 for an example.

__________ 1. Folding takes place when a sudden displacement occurs along a fault.

__________ 2. The ocean floors are geologically very young.

__________ 3. Geologists reconstruct past events in the earth's history in terms of geologic processes still at work today.

__________ 4. An igneous rock is older than the youngest bed it intrudes.

__________ 5. Radioactive dating and fossil identification are two methods used to figure out the worldwide sequence of events that have shaped the earth's surface.

__________ 6. Radiocarbon is found in all living organisms.

__________ 7. According to the fossil record, organisms show a progressive change from simple forms to more complex forms.

__________ 8. Geologic processes during Precambrian time were very different from those of today.

__________ 9. The earth is about 6000 years old.

__________ 10. Most of the coal deposits in the eastern United States formed during the Jurassic Period.

Fill in the Blank

1. A ____________________ is a rock fracture along which one side has moved relative to the other.

2. Today's continents were once all part of a huge landmass called ____________________.

3. According to the theory of ____________________ (2 words), the earth's lithosphere is divided into huge slabs or plates.

4. A ____________________ fault is a type of plate boundary which occurs where the edges of two plates slide past each other.

5. The French biologist George Cuvier greatly influenced geology at the beginning of the nineteenth century by his study of ____________________.

6. According to the principle of ____________________ change, the geologic processes in operation today were in operation in the past.

7. In a sequence of undisturbed sedimentary rocks, the lowest bed is ____________________ in age than the beds above.

8. An ____________________ is a buried surface of erosion.

9. ____________________ is a beta radioactive isotope of carbon with a half life of 5700 years.

10. The most ancient rocks known are believed to be about ____________________ billion years old.

11. The times of ____________________ are used to divide geologic history into units of time called periods.

12. Periods are divided into shorter units of time called ____________________.

13. The oldest fossils of the Paleozoic Era are those of marine ____________________.

14. Dinosaurs first appeared during the ____________________ Era.

15. We currently live in the ____________________ Era.

Matching

Match the event with the period or epoch in which it occurred.

1._____ Ordovician

2._____ Silurian

3._____ Devonian

4._____ Carboniferous

5._____ Permian

6._____ Triassic

7._____ Jurassic

8._____ Cretaceous

9._____ Tertiary

10.____ Pleistocene

a. the Himalayas are uplifted
b. first mammals appear
c. dinosaurs become extinct
d. first vertebrates (fishes) appear
e. first true birds appear
f. first amphibians appear
g. the Ice Age
h. first land plants
i. reptiles become dominant
j. coal-forming swamps in eastern North America and Europe

WEB LINKS

View reconstructions of tectonic plates over geologic time at

http://www.ig.utexas.edu/research/projects/plates/plates.html

Take a virtual time machine to see what the earth looked like in the past and what it might look like in the future at

http://www.scotese.com/earth.htm

Take an interactive tour of geologic time at

http://www.ucmp.berkeley.edu/help/timeform.html

Watch the world's population grow at

http://math.berkeley.edu/~galen/popclk.html

ANSWER KEY

Multiple Choice

1. b 2. a 3. c 4. d 5. d 6. c 7. a 8. a 9. c 10. b 11. a 12. d 13. b 14. b 15. c 16. d 17. b 18. b 19. c 20. d

True or False

1. False. Folding results from a slow, continuous movement of the earth's crust.
2. True
3. True
4. False. An igneous rock is <u>younger</u> than the youngest beds it intrudes.
5. True
6. True
7. True
8. False. Precambrian geologic processes resembled those of today.
9. False. The earth is approximately 4.5 billion years old.
10. False. Most of the coal deposits in the eastern United States formed during the <u>Carboniferous</u> Period.

Fill in the Blank

1. fault
2. Pangaea
3. plate tectonics
3. transform
5. fossils
6. uniform
7. older
8. unconformity
9. radiocarbon (or ^{14}C)
10. 4
11. extinctions
12. epochs
13. invertebrates
14. Mesozoic
15. Cenozoic

Matching

1. d 2. h 3. f 4. j 5. i 6. b 7. e 8. c 9. a 10. g

Chapter 16

THE SOLAR SYSTEM

OUTLINE

GOALS

1. Distinguish between the rotation and revolution of a planet and state the two regularities of these motions shared by most planets and satellites.
2. Identify the inner and outer planets and list the common properties of the members of each group.
3. Discuss why Pluto does not fit into either of the above groups.
4. Discuss the nature and appearance in the sky of comets.
5. Give the two causes of the deflection of comet tails so they always point away from the sun.
6. Describe the appearance in the sky and the nature of meteors.
7. Describe how Venus is environmentally different from earth.
8. Discuss the possibility of life on Mars.
9. State what asteroids are and describe the danger asteroids pose.
10. Describe the Great Red Spot of Jupiter.
11. Discuss the nature of Saturn's rings and explain how this was discovered before spacecraft visited Saturn.
12. Use a diagram to account for the phases of the moon.
13. Explain why eclipses of the sun and moon occur.
14. Describe the surface features of the moon.
15. Outline the evolution of the moon's surface.
16. Discuss the various theories of the origin of the moon.

CHAPTER SUMMARY

Chapter 16 is a tour of the **solar system**. **Asteroids**, **comets**, and **meteors** are described in terms of their composition and origin. The individual planets of the solar system are described in terms of their planetary motions, physical features, and environmental conditions. The **phases** of the **moon**, and **solar** and **lunar eclipses** are discussed. Features of the lunar surface are described, and theories concerning the moon's origin are presented.

CHAPTER OUTLINE

16-1. The Solar System

A. The **solar system** consists of the sun, the planets, their **satellites** (moons), and other smaller bodies such as **asteroids**, comets, and meteoroids.

B. The asteroids are small objects that follow separate orbits between Mars and Jupiter.

C. Planets **revolve** around the sun and **rotate** on their axes.

1. All revolutions and rotations, except for Venus and a few minor satellites, are in the same direction. Uranus rotates about an axis only 8° from the plane of its orbit.
2. All the orbits except those of Pluto and the comets lie nearly in the same plane.

D. The **inner planets** are Mercury, Venus, Earth, and Mars.

1. The inner planets are relatively small and rocky, have similar densities, have low escape speeds, and rotate fairly slowly on their axes.
2. Of the inner planet satellites, the earth's moon is the only satellite of any size.

E. The **outer planets** are Jupiter, Saturn, Uranus, Neptune, and Pluto.

1. All the outer planets except Pluto are large, not very dense, rotate fairly rapidly, have high escape speeds, and are composed mostly of liquefied gases (chiefly hydrogen and helium).
2. The outer planets have a total of at least 130 satellites.
3. Pluto is a small planet composed of rock and frozen gases.

16-2. Comets

A. **Comets** are leftover matter from the early history of the solar system and consist mainly of ice and dust.

B. Most comets move in long, narrow elliptical orbits about the sun.

C. Comets are visible only when close to the sun partly because of sunlight scattered by cometary material but mainly through the excitation of vaporized gases by solar ultraviolet radiation.

D. A comet's tail always points away from the sun because of pressure from the sun's radiation on the comet's dust particles and from the **solar wind** (the stream of ions flowing outward from the sun) sweeping the comet's gases with it.

16-3. Meteors

A. **Meteoroids** are small fragments of matter, formed as a result of asteroid collisions and from comet debris, that the earth meets as it travels through space.
B. **Meteors**, or "shooting stars," are the glows of meteoroids as they pass through the earth's atmosphere.
C. **Meteorites** are meteoroids that have fallen to earth.
D. **Micrometeorites** are fine, dustlike meteorites that float through the earth's surface without burning up.
E. Meteor showers occur at specific times of the year when the earth moves through a swarm of meteoroids that all follow the same orbit about the sun.
F. Most meteorites fall into two classes:
    1. Stony meteorites, which consist of silicate minerals.
    2. Iron meteorites, which consist largely of iron and some nickel.
G. The sizes and rate of arrival of meteorites bombarding earth and earth's neighbors in space fell to present rates about 3 billion years ago.

16-4. Mercury

A. Mercury is the planet closest to the sun.
B. Mercury takes 59 earth days to complete one rotation.
C. Mercury has almost no atmosphere. Temperatures fall to -180°C on the dark side and reach 425°C on the sunlit side.
D. The surface of Mercury is pocked by meteoroid craters.

16-5. Venus

A. Venus is the brightest object in the sky apart from the sun and moon.
B. Venus rotates clockwise (looking down from above its north pole), unlike the earth and the other planets which rotate counterclockwise.
C. A day on Venus is 243 earth days long because of the planet's slow rotation.
D. Venus resembles the earth in size and shape; however, the atmospheric pressure is about 90 times that of earth, and the surface temperature averages over 400°C. Under such extreme conditions, life on Venus seems impossible.

16-6. Mars

A. Mars is second only to Venus in brightness and is reddish orange in color.
B. Mars's satellites, Phobos and Deimos, discovered by Asaph Hall in 1877, are very small.
C. Mars's diameter is slightly over half that of the earth, and its mass is 11 percent of the earth's mass.
D. Mars probably has about the same composition as the earth.

E. Mars's period of rotation is a little over 24 h; its period of revolution requires nearly two earth years.
F. The Martian day and night are about the same length as earth's; however, the seasons are 6 months long.

16-7. Is There Life on Mars?

A. Conditions on Mars are extremely harsh.
   1. Mars receives much less solar energy than does earth.
   2. Mars's atmosphere is extremely thin and consists largely of carbon dioxide.
   3. Mars's atmosphere retains very little heat after nightfall and is unable to screen out harmful solar ultraviolet radiation.
   4. Liquid water is scarce on Mars, although it might have once been abundant.

B. Despite the harsh conditions, life of some kind is conceivable on Mars, although no evidence of present or past life has been detected in photographs or by the Viking spacecraft that landed on Mars in 1976.
C. Conditions suitable for life possibly once existed on Mars and may do so today.

16-8. Asteroids

A. Asteroids are small, rocky bits of matter from the early solar system that never became part of larger bodies.
B. Most asteroids orbit the sun in a belt between Mars and Jupiter.
C. Ceres, the largest asteroid and the first to be identified, was discovered by the Italian astronomer Giuseppe Piazzi.
D. Several thousand asteroids have been tracked and named.
E. Asteroids have collided with earth in the past and will do so again..

16-9. Jupiter

A. Jupiter is the largest planet.
B. The Great Red Spot is a huge atmospheric disturbance. It is probably the same spot described by the French astronomer Cassini who used it to determine Jupiter's period of rotation.
C. Jupiter's period of rotation is a little less than 10 h.
D. Jupiter has 61 known satellites; the largest is as big as Mercury, the smallest about 2 km in diameter.
   1. All of Jupiter's moons, except Callisto, show signs of geological activity.
   2. The four outermost satellites revolve "backward" around Jupiter.
   3. Jupiter has a faint ring.

E. Jupiter's volume is about 1300 times that of the earth, but its mass is only 318 times as great.
F. Jupiter has a magnetic field nearly 10 times stronger than the earth's.

G. Jupiter radiates over twice as much energy as it receives from the sun.

16-10. Saturn

A. Saturn is much like Jupiter in many respects, though smaller and less massive.
B. There are at least 18 satellites of Saturn, the largest being Titan, which is the only satellite in the solar system with an atmosphere.
C. A number of **rings** surround Saturn at its equator.
   1. The rings consist of numerous small bodies each of which revolves around Saturn as a miniature satellite.
   2. In 1980 and 1981 Voyager spacecraft reached Saturn and reported that the bodies that compose the rings are chunks of rock and ice.

16-11. Uranus, Neptune, Pluto

A. Uranus was discovered in 1781, Neptune in 1846, and Pluto in 1930.
B. Uranus rotates about an axis only 8° from the plane of its orbit.
C. The Voyager 2 spacecraft reported the presence of rings around Uranus and Neptune.
D. Neptune's atmosphere is more stormy than that of Uranus.
E. Pluto seems to be about two-thirds the size of the moon and to consist of rock, ice, and frozen methane.
F. Pluto's satellite, Charon, has a diameter half that of Pluto and might have an ocean of liquid water under a covering of ice.

16-12. Phases

A. The moon is earth's satellite.
   1. The distance from the earth to the moon averages 384,000 km.
   2. The moon has a diameter of 3476 km.
   3. The moon circles the earth every 27⅓ days.
   4. Since the moon's rotation keeps pace exactly with its revolution of the earth, the same hemisphere of the moon always faces the earth.
   5. The light from the moon is reflected sunlight.

B. Relative to the stars, the moon's orbital period is 27⅓ days; relative to the sun, it is 29 ½ days.
C. During each 29½-day period the moon goes through its cycle of **phases**.
   1. The phases represent the amount of the moon's illuminated surface visible to us in different parts of its orbit.
   2. A full moon is on the opposite side of the earth from the sun, so the side facing earth is fully illuminated.
   3. A new moon is moving approximately between us and the sun, so the side facing earth is in shadow.

D. The new moon is not completely dark because of **earthshine**—sunlight reflected from the earth's surface that reaches the moon

16-13. Eclipses

A. **Eclipses** occur when the moon passes exactly between the earth and the sun and obscures the sun (solar eclipse) or when the moon passes exactly behind the earth so that earth's shadow obscures the moon (lunar eclipse).
B. The circular shape of the earth's shadow during a lunar eclipse is evidence for earth's spherical shape.
C. Partial eclipses of the sun take place when the moon is not exactly aligned with the sun, so that only part of the sun's disk is obscured. During a total eclipse, the moon completely blocks out light from the sun.
D. An **annular** eclipse of the sun occurs when the moon is farthest from the earth and its apparent diameter is less than that of the sun. The result is a ring of sunlight appearing around the rim of the moon.

16-14. The Lunar Surface

A. The moon has no atmosphere and no surface water.
B. The two main kinds of lunar landscape include:
    1. The dark, relatively smooth **maria** (the singular is **mare**)
    2. The lighter, ruggedly mountainous highlands
C. The maria consist of lava flows that have been broken up by meteoroid impacts.
D. The lunar highlands are scarred by numerous craters.
    1. Certain craters have conspicuous streaks of light-colored matter radiating outward called **rays**.
    2. The rays consist of lunar material sprayed outward after meteoroid impacts.
E. Lunar **rilles** of the highlands are narrow channels that look like dried-up riverbeds and were probably created by the collapse of subsurface channels through which lava once flowed.
F. Seismographic data reveal that the moon has a rigid crust, a thick solid mantle, and a small, dense core.

16-15. Evolution of the Lunar Landscape

A. The analysis of lunar rock and soil samples has led to a number of conclusions about the history of the moon.
    1. Some lunar rocks solidified soon after the solar system came into being.
    2. The moon's landscape was shaped by meteoroid bombardment and volcanic activity.
B. The youngest rocks found on the moon are 3 billion years old, indicating that all igneous activity stopped then.

C. Meteoroids continue to crater the lunar landscape.

16-16. Origin of the Moon

A. Until recently theories of the moon fell into three categories:
   1. The moon split away from the earth.
   2. The moon was formed elsewhere and later captured by the earth's gravitational field.
   3. The moon and the earth came into being together as a double-planet system.

B. All three theories have serious weaknesses.

C. A fourth proposal, the collision hypothesis, is widely accepted.
   1. The collision hypothesis suggests that another planet, a little larger than Mars and with a slightly different composition from that of the earth, crashed into the earth.
   2. The mantle of the other planet and some of the earth's mantle were thrown off into orbit to form the moon.
   3. The other planet's iron core was added to the earth's core.

KEY TERMS AND CONCEPTS

The questions in this section will help you review the key terms and concepts from Chapter 16.

Multiple Choice

Circle the best answer for each question.

1. With his telescope, Galileo discovered
   a. the lunar maria
   b. four moons of Jupiter
   c. Neptune
   d. the asteroids

2. Compared to the orbits of the planets, the orbits of comets about the sun are
   a. perfect circles
   b. longer and narrower
   c. shorter and wider
   d. exactly the same as those of the planets

3. Meteorites
   a. are impossible to distinguish from terrestrial rocks
   b. are composed chiefly of ice and dust
   c. are the flashes of light that meteoroids produce when they enter the earth's atmosphere
   d. reach the earth's surface without burning up

4. Mercury and Venus are similar in that they
   a. rotate clockwise about their axes
   b. possess thick atmospheres
   c. lack moons
   d. give off over twice as much energy as they receive from the sun

5. The first object of human origin to leave the solar system permanently is the
   a. Pioneer 10 spacecraft
   b. Mariner 10 spacecraft
   c. Viking spacecraft
   d. Galileo spacecraft

6. The orbits of all planets lie nearly in the same plane except for that of
   a. Venus
   b. Uranus
   c. Mars
   d. Pluto

7. Which one of the following statements about the inner planets is incorrect?
   a. The inner planets are relatively small.
   b. They are composed largely of rocky material.
   c. They rotate fairly rapidly on their axes.
   d. They possess low escape speeds.

8. Spacecraft that have landed on Venus found
   a. extensive seas
   b. an atmosphere rich in oxygen
   c. surface temperatures hot enough to melt lead
   d. thick clouds composed mainly of water

9. The seasons on Mars
   a. last little over 24 hours
   b. last nearly 2 years
   c. are 6 months long
   d. are about the same lengths as the earth's seasons

10. Life on Mars, if it exists, would have to adapt to all the following conditions except
    a. intense solar ultraviolet radiation
    b. an atmosphere rich in oxygen
    c. an environment low in liquid water
    d. a low average surface temperature

11. Ceres is
    a. an asteroid
    b. one of Mars's moons
    c. Pluto's moon
    d. the largest satellite of the solar system

12. Jupiter's moon Europa
    a. has active volcanism
    b. is Jupiter's largest moon
    c. has a faint ring
    d. appears to have an ocean of liquid water beneath an ice surface

13. The rings of Saturn
    a. are split into thousands of narrow ringlets
    b. consist of very large captured asteroids
    c. are composed mainly of gaseous hydrogen and helium
    d. are composed of metallic hydrogen

14. An annular eclipse of the sun occurs when the
    a. earth is between the moon and the sun
    b. earth is farthest from the sun
    c. moon passes above or below the direct line from the sun to the earth
    d. moon is farthest from the earth

15. Examination of lunar rocks indicates that all igneous activity on the moon stopped about
    a. 20 million years ago
    b. 1.5 billion years ago
    c. 3 billion years ago
    d. 4.6 billion years ago

True or False

Decide whether each statement is true or false. If false, either briefly state why it is false or correct the statement to make it true. See Chapter 1 or 2 for an example.

__________ 1. The same hemisphere of the moon always faces the earth because the moon does not rotate about its axis.

__________ 2. A solar eclipse occurs when the moon passes exactly between the earth and the sun.

_________ 3. The planet most like the earth in size and shape is Mars.

_________ 4. Like the earth, the surface of Mars is broken into a number of crustal plates.

_________ 5. The asteroids are probably similar in composition to comets.

_________ 6. Jupiter consists mainly of hydrogen and helium.

_________ 7. Saturn's rings are chunks of rock and ice.

_________ 8. The Great Dark Spot is a huge atmospheric disturbance found on the planet Jupiter.

_________ 9. The lunar maria are "seas" of frozen water.

_________ 10. The most likely explanation for the origin of the moon states that the moon split away from the earth.

Fill in the Blank

1. The ____________________ (2 words) consists of the sun together with the accompanying planets, their satellites, ans other smaller bodies.

2. The ____________________ are small, rocky bits of matter that follow separate orbits between Mars and Jupiter.

3. Planets ____________________ around the sun and ____________________ on their axes.

4. The ____________________ planets are Mercury, Venus, Earth, and Mars; the ____________________ planets are Jupiter, Saturn, Uranus, Neptune, and Pluto.

5. ____________________ are leftover matter from the solar system's early history and consist mainly of dust and ice (frozen gases).

6. The tail of a comet always points away from the ____________________.

7. The ___________________ (2 words) is a stream of ions that constantly flows outward from the sun.

8. ___________________ are meteoroids that have fallen to earth.

9. The ___________________ (3 words) is a huge atmospheric disturbance on the planet Jupiter.

10. The planet ___________________ is the brightest object in the sky apart from the moon and sun.

11. The moon goes through its cycle of ___________________ during each 29½-day period.

12. During an ___________________ eclipse of the sun, the moon is farthest from the earth and its apparent diameter is less than that of the sun.

13. ___________________ consist of light-colored material radiating outward from lunar craters.

14. The ___________________ of the lunar highlands were probably created by the collapse of subsurface channels through which lava once flowed.

15. The planet whose orbit is most elliptical in shape is ___________________.

Matching

1._____	Mercury	a.	its moon is named Charon
		b.	planet closest to the sun
2._____	Venus	c.	largest planet
		d.	planet with the most distinctive rings
3._____	Titan	e.	the red planet
		f.	only planet that rotates clockwise on its axis
4._____	Mars	g.	only satellite having an atmosphere
5._____	Io	h.	satellite of Jupiter having active volcanoes
6._____	Jupiter	i.	rotates about an axis only 8° from the plane of its orbit
7._____	Saturn	j.	largest asteroid
8._____	Uranus		
9._____	Pluto		
10.____	Ceres		

WEB LINKS

Take a multimedia tour of the solar system at

http://seds.lpl.arizona.edu/billa/tnp/

ANSWER KEY

Multiple Choice

1. b 2. b 3. d 4. c 5. a 6. d 7. c 8. c 9. c 10. b 11. a 12. d 13. a 14. d 15. c

True or False

1. False. The same hemisphere of the moon always faces the earth because the moon rotates on its axis once during each revolution of the earth.
2. True
3. False. The planet most like earth in size and shape is Venus.
4. False. The Martian surface seems to consist of an unbroken shell of rock. Mars lacks sufficient internal heat to drive tectonic processes like those of earth.
5. False. The asteroids are probably similar in composition to stony and stony-iron meteorites.
6. True
7. True
8. False. The Great Dark Spot is a huge atmospheric disturbance found on the planet Neptune.
9. False. The lunar maria consist of lava flows pulverized by meteoroids.
10. False. The fragmentation theory of the moon's origin has serious weaknesses. A more attractive hypothesis suggests that the moon was formed by the collision of another planet with the earth.

Fill in the Blank

1. solar system
2. asteroids
3. revolve, rotate
4. inner, outer
5. comets
6. sun
7. solar wind
8. meteorites
9. Great Red Spot
10. Venus
11. phases
12. annular
13. rays
14. rills
15. Pluto

Matching

1. b 2. f 3. g 4. e 5. h 6. c 7. d 8. i 9. a 10. j

Chapter 17
THE STARS

OUTLINE

GOALS

1. Describe how analyzing the spectrum of a star can provide information on the star's structure, temperature, composition, condition of matter, magnetism, and motion.
2. Sate what is meant by the photosphere of the sun.
3. Explain why solar energy cannot come from combustion.
4. Identify the basic process that gives rise to solar energy.
5. Describe how the elements more massive than hydrogen are created and distributed throughout the universe.
6. Describe the appearance in the sky and the origin of auroras.
7. Discuss sunspots, the sunspot cycle, and some effects on the earth that are correlated with sunspot activity.
8. Define light-year.
9. Describe the parallax method of finding the distance to a star.
10. Describe how the distance to a star can be found by comparing its apparent and intrinsic brightness.
11. Explain why Cepheid variable stars are useful in finding the distance of a star group.
12. Describe how the mass and size of a star can be found.
13. Account for the relatively small range of stellar masses.
14. State what a brown dwarf is.
15. Interpret in terms of stellar structure the observation that nearly all stars have absorption (dark line) spectra.
16. State what is plotted on a Hertzsprung-Rusell (H-R) diagram.
17. Draw an H-R diagram and indicate the positions of main-sequence stars, red giants, and white dwarfs.
18. Compare the properties of red giants and white dwarf stars.
19. Outline the life history of an average star like the sun.
20. Outline the life history of a very massive star.
21. State what a supernova is.
22. Define neutron star and pulsar and discuss the connection between them.
23. Describe what black holes are and explain how they can be detected.

CHAPTER SUMMARY

Chapter 17 introduces us to stellar astronomy and to the properties and life histories of the stars. The tools and techniques of modern stellar astronomy are described, and the importance of light as a messenger of information about the universe is stressed. The nature of the sun is discussed, and the types of nuclear reactions that take place in the sun and other stars are presented. Methods for determining stellar distances, stellar properties, and the age of the universe are discussed. The processes leading to the formation of **neutron stars, pulsars, brown dwarfs**, and **black holes** are given.

CHAPTER OUTLINE

17-1. The Telescope

A. The English astronomer William Herschel was the first to build and use a large reflecting telescope.
   1. In a reflecting telescope, light is reflected from a concave mirror instead of being refracted through a lens.
   2. All modern astronomical telescopes are reflectors.

B. The latest reflecting telescopes use multiple mirrors instead of single large mirrors.

C. Large telescopes have greater light-gathering power and better ability to **resolve** (distinguish) small details than do smaller telescopes.

D. The light collected by a telescope is directed to a photographic plate or electronic sensor which can detect objects too faint for the eye to pick up.

17-2. The Spectrometer

A. Telescopes are combined with spectrometers to collect more information about stars.

B. The spectrometer breaks light up into its separate wavelengths. The resulting spectrum is recorded on a photographic plate or electronic medium.

17-3. Spectrum Analysis

A. An **absorption spectrum** (a spectrum of dark lines on a continuous colored background) reveals that a star's structure consists of a hot interior surrounded by a cooler atmosphere. Most stars have absorption spectra.

B. The surface temperature of a star can be determined by spectrum analysis. The point of maximum wavelength intensity in the spectrum is a measure of its temperature.
   1. The hottest stars are blue-white.
   2. Stars of intermediate temperature are orange-yellow.
   3. The coolest visible stars are red.

C. The composition of a star can be found from its spectrum because each element in a star has a spectrum consisting of lines with characteristic wavelengths.

D. Identification of spectral lines can reveal something about the physical conditions in which the elements exist.
E. The **Zeeman effect** permits the detection of magnetic fields of stars.
F. The motion of a star toward or away from the earth is shown by doppler shifts in its spectral lines.

17-4. Properties of the Sun

A. The mass of the sun is 1.99×10^{30} kg.
B. The sun's radius is 6.96×10^8 m.
C. The surface temperature of the sun is 6000 K.
D. The glowing gas surface of the sun is called the **photosphere**.
E. The sun consists mainly of hydrogen and helium.
F. The elements in the sun are present as individual atoms or ions.
G. Solar **prominences** are flamelike protuberances that project from the sun's atmosphere into space. They are associated with sunspots and seem to have magnetic fields associated with them.
H. The sun's **corona** consists of ions and electrons and is visible during solar eclipses.
I. The **solar wind** is the outward flow of ions and electrons from the sun's corona.

17-5. The Aurora

A. An **aurora** is a luminous atmospheric display produced by the excitation of atmospheric gases by streams of fast protons and electrons from the sun.
   1. The **aurora borealis** is the name given to this phenomenon in the northern hemisphere, and **aurora australis** in the southern.
   2. Auroras are most common in the far north and far south.
B. **Airglow** is the faint glow in the night sky due to less concentrated streams of solar particles interacting with the upper atmosphere. The degree of brightness varies with solar activity.

17-6. Sunspots

A. **Sunspots** are cooler areas on the solar surface that appear dark only by comparison with the brighter solar surface around them.
   1. Sunspots change continually in form and have lifetimes of from 2 to 3 days to more than a month.
   2. Galileo associated the movement of sunspots across the sun's disk with solar rotation.
   3. Sunspots generally appear in groups.
   4. Strong magnetic fields are associated with sunspots.
   5. The sunspot cycle is about 11 years long.

B. A number of effects observable on earth are associated with the sunspot cycle. These include:
   1. **Solar storms** (disturbances in the terrestrial magnetic field)
   2. Shortwave radio fadeouts
   3. Changes in cosmic-ray intensity
   4. Unusual auroral activity
   5. Some aspects of weather and climate

17-7. Solar Energy

A. The sun's temperature and pressure at its center are estimated to be 14 million K and 1 billion atm, respectively.
B. Under these conditions, matter in the sun's interior consists of free electrons and positive nuclei surrounded by a few electrons or none at all.
C. These atomic fragments move so rapidly that two atomic nuclei, despite their repulsive electric force, can fuse to form a single large nucleus.
D. When this occurs, the new nucleus has a little less mass than the combined masses of the reacting nuclei. This missing mass is converted to energy.
E. Most solar energy comes from the conversion of hydrogen into helium in nuclear fusion reactions. This takes place directly by collisions of hydrogen nuclei (proton-proton cycle) and indirectly by a series of steps in which carbon nuclei absorb a succession of hydrogen nuclei (carbon cycle).
F. The sun converts more than 4 billion kg of matter into energy every second and has emitted energy at a steady rate for a long time.
G. The heavier elements are formed from hydrogen and helium as raw materials subjected to extreme conditions of high temperatures and pressures.

17-8. Stellar Distances

A. In 1838, the German astronomer Friedrick Bessel used the shift in the relative position of a star to make direct measurements of distances to the nearer stars.
B. A **parallax** is the apparent shift in a star's position.
C. A **light-year** is the distance light travels in a year and is equal to 9.46×10^{12} km.
D. The **apparent brightness** of a star is its brightness as seen from the earth; its **intrinsic brightness** is the star's true brightness.
   1. The apparent brightness of a star depends on its intrinsic brightness and its distance from the earth.
   2. If both the apparent and the intrinsic brightness of a star are known, its distance from the earth can be calculated.
E. The American astronomer Walter Adams discovered a way to find the intrinsic brightness of a star by examining its spectrum.

17-9. Variable Stars

A. A **variable star** is one whose brightness varies continually.
B. Most variable stars repeat a fairly definite cycle of change; others show irregular fluctuations.
C. **Cepheid variables** are a special class of variable stars that are useful to astronomers in determining the distance to certain star groups.
   1. Cepheid variables are very bright yellowish stars 5 to 10 times as heavy as the sun.
   2. The American astronomer Henrietta Leavitt discovered that the intrinsic brightness of a Cepheid variable is related to its period of fluctuation.
   3. Comparing the intrinsic brightness with the Cepheid variable's apparent brightness gives its distance.

17-10. Stellar Motions

A. The stars are not fixed in space; most stars are moving at speeds of several kilometers per second relative to the earth.
B. The sun and the planets are moving toward the constellation Cygnus at a speed of 200 to 300 km/s.

17-11. Stellar Properties

A. Only the masses of **binary stars** can be determined, but such stars are common.
B. Stellar masses range from 1/40 to 100 times that of the sun.
C. The majority of stars have surface temperatures between 3000 and 12,000 K.
D. The size of a star can be found from its temperature and its intrinsic brightness.
E. The diameters of the smallest stars are about 20 km across; the largest have diameters 500 or more times that of the sun.

17-12. H-R Diagram

A. **A Hertzsprung-Russell** (or **H-R**) **diagram** is a plot of intrinsic brightness versus temperature for stars.
   1. Each point on the diagram represents a particular star.
   2. About 90 percent of all stars belong to the **main sequence**.
   3. The most abundant stars in the main sequence are **red dwarfs**.
   4. Outside of the main sequence, most of the remaining stars belong to the **red giant** class and to the **white dwarf** class.
B. The position of a star on the H-R diagram is related to its physical properties.
C. The sun occupies a middle position on the H-R diagram. Such stars have moderate temperatures, densities, and masses, rather small diameters, and spectra in which lines

of metallic elements are prominent.

D. Red giants have low densities, large diameters, low surface temperatures, and hot cores.
E. White dwarfs have small diameters (they are comparable in size to the earth), high densities, and high surface temperatures.
F. The density of a white dwarf is about 10^6 g/cm^3.
G. To attain densities this high, the matter in white dwarfs must consist of collapsed atoms whose electrons and nuclei are packed closely together.
H. Although the universe contains numerous white dwarfs, only a few thousand are known, because they are very faint and only the nearer ones can be seen.

17-13. Stellar Evolution

A. Stars originate in gas clouds in space. These clouds are largely hydrogen.
B. Gravitational contraction results in a heated, glowing clump of matter.
C. Some thousands or millions of years later the star's temperature rises to the point where nuclear reactions begin.
D. The greater the mass of an H-R main sequence star, the higher its temperature.
E. When the hydrogen supply runs low in a star like the sun, several events take place.
   1. Gravitational contraction increases core temperature and new nuclear reactions become possible.
   2. The star becomes a red giant.
   3. The new nuclear reactions run out of fuel and the star shrinks to form a white dwarf.
   4. At this stage, a bubblelike shell of gas from the outer part of the star, called a **planetary nebula**, is released into space.
   5. Ultimately the star ceases to radiate at all and becomes a **black dwarf**.

17-14. Supernovas

A. A star having over 8 times the sun's mass at the upper end of the H-R main sequence collapses when its fuel runs out. It then explodes violently, creating a **supernova**.
B. A supernova is billions of times brighter than the original star.
C. Nuclear reactions during the explosion create the heaviest elements and send them into space.
D. After the supernova explosion, the remaining matter is compressed into a dwarf star of extraordinary density called a **neutron star**.

17-15. Pulsars

A. A **pulsar** is a rapidly spinning neutron star that emits bursts of light and radio waves.
B. The power output of a pulsar is about 10^{26} W.

17-16. Black Holes

A. A **black hole** is a "dead" star whose matter is packed so densely that its gravitational field is strong enough for the escape speed to be greater than the speed of light.
B. Only extremely heavy stars become black holes; lighter stars eventually become white dwarfs or neutron stars.
C. If a black hole is a member of a double-star system, its presence can be revealed by its gravitational pull on the other star and by the emission of x-rays as its gravitational field pulls matter away from the other star.
D. Enormous black holes are thought to be at the centers of galaxies.

KEY TERMS AND CONCEPTS

The questions in this section will help you review the key terms and concepts from Chapter 17.

Multiple Choice

Circle the best answer for each question.

1. Astronomers have a great deal of detailed information on thousands of stars because
   a. of unmanned space probes to these stars
   b. the stars are easily seen using the largest telescopes
   c. all stars are exactly like the sun
   d. of spectroscopic analysis

2. The latest generation of telescopes
   a. utilize a single, large lens
   b. utilize a number of individual mirrors linked to produce a single image
   c. consist of a single, large mirror and a series of magnifying lenses
   d. refract light through a lens

3. The color of a star depends on its
   a. composition
   b. temperature
   c. period of rotation
   d. distance from the earth

4. The composition of a star can be determined by examining its
   a. color
   b. magnetic field
   c. spectrum
   d. sunspots

5. A star moving toward the earth has a spectrum whose lines are
   a. continuous with no gaps in the band of color
   b. blurred
   c. doppler-shifted toward the violet end of the spectrum
   d. doppler-shifted toward the red end of the spectrum

6. The sun consists largely of
   a. hydrogen and helium
   b. oxygen and nitrogen
   c. carbon and sodium
   d. neon and argon

7. Which one of the following statements about sunspots is correct?
   a. The sunspot cycle is about 4 years long.
   b. Magnetic fields are associated with sunspots.
   c. Sunspots appear as bright patches on the sun's surface.
   d. Sunspots are usually clustered near the sun's equator.

8. The American astronomer Walter Adams discovered a way to determine the intrinsic brightness of a star by examining the star's
   a. distance from the earth
   b. sunspot activity
   c. spectrum
   d. size

9. The American astronomer Henrietta Leavitt discovered that the intrinsic brightness of a Cepheid variable is related to its
   a. period of fluctuation
   b. color
   c. period of rotation
   d. doppler shift

10. Which statement best describes a white dwarf?
    a. High mass and volume, low density
    b. Cool surface, high total radiation
    c. Hot surface, low density, small size
    d. Hot surface, low total radiation, high density

11. The sun is
    a. at the beginning of its lifetime in the main sequence
    b. near the end of its lifetime in the main sequence
    c. about halfway through its lifetime in the main sequence
    d. about ready to begin its lifetime as a main sequence star

12. Pulsars
    a. have a total energy output much less than that of the sun
    b. are rapidly spinning black holes
    c. have weak magnetic fields
    d. have very small diameters

True or False

Decide whether each statement is true or false. If false, briefly state why it is false or correct the statement to make it true. See Chapters 1 or 2 for an example.

__________ 1. All modern astronomical telescopes are reflecting telescopes.

__________ 2. One advantage of large telescopes used in stellar astronomy is their increased magnification of distant stars.

__________ 3. Because most stars exhibit an absorption spectrum, we know that most stars have hot interiors surrounded by cooler atmospheres.

__________ 4. The hottest stars are red giants.

__________ 5. The proton-proton cycle is the chief nuclear reaction sequence that takes place in the sun.

__________ 6. Sunspots are responsible for observable effects on the earth such as shortwave radio fadeouts, disturbances in the magnetic field, and unusual auroral activity.

__________ 7. The number of sunspots increases and decreases with time in a regular cycle of 27 years.

__________ 8. If both the apparent brightness of a star and the intrinsic brightness of a star are known, we can calculate its distance.

__________ 9. The greater the mass of a main sequence star, the lower its temperature.

__________ 10. Of all the stars belonging to the main sequence, red dwarf stars are the least abundant.

Fill in the Blank

1. An ____________________ spectrum is produced when light from a hot object passes through a cooler gas.

2. The ____________________ wind is the outward flow of ions and electrons from the sun's corona.

3. The aurora ____________________ in the northern hemisphere, and the aurora ____________________ in the southern hemisphere, are caused by streams of fast protons and electrons in the solar wind that excite gases in the upper atmosphere to emit light.

4. ____________________ is faint glow in the night sky caused by less concentrated stream of solar particles that interact with the upper atmosphere.

5. ____________________ storms are disturbances in the earth's magnetic field caused by vast streams of energetic protons ans electrons emitted by the sun from the vicinity of sunspot groups.

6. The net result of the proton-proton cycle is the combination of four hydrogen nuclei to form a ____________________ nucleus and two positrons.

7. A ____________________ is the distance light travels in a year.

8. The ____________________ brightness of a star is the brightness of a star as seen from earth; its ____________________ brightness is its true brightness.

9. ____________________ variables are fairly old, bright yellow stars 5 to 10 times as massive as the sun whose related intrinsic brightness and periods of variation permit their distances to be found.

10. The ____________________ diagram is a graph on which the intrinsic brightnesses of stars are plotted versus their temperatures.

11. The sun is a ____________________ sequence star on the H-R diagram.

12. ____________________ (2 words) stars have small diameters, high densities, and high surface temperatures.

13. ____________________ (2 words) stars have large diameters, low densities, and low surface temperatures.

14. A ____________________ is a rapidly spinning neutron star that emits bursts of light and radio waves.

15. At the very end of the sun's life history, it will become a ____________________ (2 words).

Matching

1._____ sunspots

2._____ prominences

3._____ parallax

4._____ variable star

5._____ photosphere

6._____ neutron star

7._____ supernova

8._____ black hole

9._____ planetary nebula

10.____ corona

a. apparent shift in a star's position
b. vast cloud of hot, rarefied gases that surrounds the sun
c. ejected shell of gas
d. glowing gas surface of the sun
e. a "dead" star whose gravitational field is so strong that not even light can escape
f. heavy star that explodes
g. star whose brightness varies continually
h. dark patches on the sun's surface
i. what is left after a supernova
j. flamelike solar protuberances

WEB LINKS

Take a virtual tour of the Big Bear Solar Observatory at

http://www.bbso.njit.edu/

ANSWER KEY

Multiple Choice

1. d 2. b 3. b 4. c 5. c 6. a 7. b 8. c 9. a 10. d 11. c 12. d

True or False

1. True
2. False. The advantage of large telescopes used in stellar astronomy is in their light-gathering power.
3. True
4. False. The hottest stars are blue-white stars.
5. True
6. True
7. False. The number of sunspots increases and decreases with time in a regular cycle of 11 years.
8. True
9. False. The greater the mass of a main sequence star, the <u>higher</u> its temperature.
10. False. Of all the stars belonging to the main sequence, red dwarf stars are the most abundant.

Fill in the Blank

1. absorption
2. solar
3. borealis, australis
4. airglow
5. solar
6. helium
7. light-year
8. apparent, intrinsic
9. Cepheid
10. Hertzsprung-Russell (or H-R)
11. main
12. white dwarf
13. red giant
14. pulsar
15. black dwarf

Matching

1. h 2. j 3. a 4. g 5. d 6. i 7. f 8. e 9. c 10. b

Chapter 18
THE UNIVERSE

OUTLINE

Galaxies
18.1 The Milky Way
18.2 Stellar Populations
18.3 Radio Astronomy
18.4 Galaxies
18.5 Cosmic Rays

The Expanding Universe
18.6 Red Shifts
18.7 Quasars

Evolution of the Universe
18.8 Dating the Universe
18.9 After the Big Bang
18.10 Origin of the Solar System

Extraterrestrial Life
18.11 Extrasolar Planets
18.12 Interstellar Travel
18.13 Interstellar Communication

GOALS

1. Describe the Milky Way galaxy and indicate the sun's location in it.
2. Compare Population I and II stars.
3. Explain what a radio telescope is.
4. List the three ways in which radio waves from space are produced.
5. Distinguish between primary and secondary cosmic waves.
6. Explain what red shifts in galactic spectra indicate about the motions of galaxies.
7. State Hubble's law and use it as evidence for the expansion of the universe.
8. Outline the properties of quasars and what they suggest about the nature of these objects.
9. Discuss the big bang theory of the origin of the universe.
10. Explain what is meant by the big crunch and why the average density of the universe is such an important quantity.
11. Outline the chief events after the big bang occurred.
12. Explain the significance of the uniform sea of radio waves that fills the universe.
13. Outline the origin of the earth.
14. Discuss why other planetary systems are hard to find.
15. Compare the likelihood of interstellar travel with that of interstellar communication.

CHAPTER SUMMARY

Chapter 18 explores the origin, evolution, and nature of the physical universe and discusses phenomena associated with the universe such as **galaxies**, **galactic nebulas**, and **quasars**. The Milky Way galaxy is described in terms of its shape, dimensions, and history. The value of radio astronomy as a tool for learning about the universe is stressed. Evidence for an expanding universe is presented, and theories concerning the evolution of the universe are discussed. The origin of the solar system and the earth are presented. The existence of extraterrestrial life and the possibilities of interstellar travel and communication are discussed.

CHAPTER OUTLINE

18-1. The Milky Way

A. A **galaxy** is a collection of stars, all held together by mutual gravitation.
B. The **Milky Way** is that portion of our galaxy visible to us in the night sky.
C. Our galaxy is shaped like a disk and is about 100,000 light-years in diameter and 10,000 light-years thick near the center.
D. Our galaxy is composed of about 100 billion stars, almost all of which revolve about its center.
E. The sun is located about 28,000 light-years from the center of our galaxy and makes a complete circuit of the galaxy once every 240 million years.
F. The center of our galaxy is obscured by clouds of gas and dust; a black hole likely exists at the galaxy's center.

18-2. Stellar Populations

A. The stars in our galaxy fall into two categories.
  1. **Population I** stars are in the galactic disk and are of all ages.
  2. **Population II** stars are in **globular clusters** above and below the galactic disk.

B. The stars in globular clusters:
  1. Range in number from about 10,000 to more than 50,000 stars
  2. Are mostly very old
  3. Are richer in hydrogen and helium and poorer in heavy elements than Population I stars

18-3. Radio Astronomy

A. A **radio telescope** is a directional antenna connected to a sensitive radio receiver.
B. Sources of radio waves from space include:
  1. The thermal motion of ions and electrons in a very hot gas
  2. High-speed electrons that move in a magnetic field
  3. Hydrogen atoms and molecules of various kinds

C. Molecules in space can be identified by their radio emissions.

18-4. Galaxies

A. Most galaxies appear as flat spirals with two curving arms that radiate from a bright nucleus.
B. Galaxies are not evenly distributed in space, but are concentrated in groups of up to a few hundred.
C. Our galaxy and about two dozen others are members of the **Local Group**.
D. The Milky Way and Andromeda galaxies will collide in 2 or 3 billion years.
E. About seven-eights of the matter in the universe is non-luminous **dark matter**, the nature of which is unknown..

18-5. Cosmic Rays

A. The Austrian physicist Victor Hess suggested that ionizing radiation from outside the earth is continually bombarding the atmosphere. This radiation was later called **cosmic radiation.**
C. Primary cosmic rays, which are the rays as they travel through space before they reach the earth, are atomic nuclei. Most were probably shot into space during supernova explosions in our galaxy.
D. Secondary cosmic rays are formed when primary cosmic waves enter the earth's atmosphere and disrupt atoms in their path to produce showers of secondary particles.

18-6. Red Shifts

A. The spectra of galaxies show red shifts, implying that all galaxies (except a few in the Local Group) are receding from the earth.
B. The amount of shift increases with the distance of the galaxy from the earth.
C. The recession speeds of galaxies can be computed from the extent of their red shifts.
D. **Hubble's law** relates the red shift of a galaxy to its distance from the earth.
    1. The greater the distance, the faster the galaxy is traveling.
    2. The speed increases by about 21 km/s per million light-years.
E. All galaxies appear to be moving away from each other; therefore, the universe seems to be expanding.

18-7. Quasars

A. **Quasars** are strong emitters of light and/or radio waves.
B. Thousands of quasars have been found and there seem to be many more.
C. Quasar red shifts are usually very large indicating that they are very distant objects.
D. Quasars radiate many times more energy than do ordinary galaxies, yet are far smaller in size.

E. Many astronomers believe that a massive black hole is possibly at the heart of every quasar, and that quasars are the cores of newly formed galaxies.

18-8. Dating the Universe

A. According to the **big bang theory**, the universe began in a violent burst in which space, time, matter, and energy came into existence.
B. Analysis of light emitted by globular clusters and white dwarf stars sets a lower limit to the age of the universe at 12-13 billion years.
B. Using Hubble's law gives a maximum age of the universe at 14 billion years.
C. The Hubble's law calculation ignores the braking effect of gravity on the expanding matter of the universe. Accounting for the effect of gravity, the age of the universe must be younger than 14 billion years.
D. Recent studies using cosmic background radiation set the age of the big bang as 13.7 billion years ago.

18-9. After the Big Bang

A. Immediately after the big bang, the universe was a compact, intensely hot mixture of matter and energy. Photons had enough energy to materialize into particle-antiparticle pairs which annihilated each other.
B. A few seconds after the big bang, the universe cooled to the point that photons no longer had enough energy to form particle-antiparticle pairs.
    1. Annihilation of particles and antiparticles continued.
    2. Because there were slightly more particles than antiparticles, there were only particles when the annihilation was finished.
C. When the universe was about a minute old, nuclear reactions began to form helium.
D. After 300,000 years, the universe cooled down to about 3000 K, cool enough for electrons and nuclei to combine into atoms.
E. Radiation that originated early in the history of the universe fills today's universe and can be detected in the range of radio waves.
F. Three observations support the big bang theory:
    1. The uniform expansion of the universe
    2. The relative abundances of hydrogen and helium
    3. The cosmic background radiation
A. The universe may be undergoing a continuous series of expansions and contractions.
B. Before the discovery that the expansion is currently speeding up, debate existed if the universe is **open** or **closed**.
    1. In an open universe, expansion will continue forever.
    2. In a closed universe, gravitational attraction will cause expansion to stop and all matter and energy will come together in a **big crunch** followed by another cycle of expansion and contraction.
    3. Recent studies, based on the discovery of dark energy, favor an open universe .

18-10. Origin of the Solar System

A. The sun and the planets formed together from a swirling cloud of gas and grains of ice and rock.
B. As the sun shrank in the process of forming a star, the resulting **protosun** left behind a spinning disk of gas and dust.
C. Bits of matter in the disk collided and stuck together to eventually form **planetesimals** that ultimately joined to become the planets.
D. Once the protosun became a true star, its intense solar wind swept the solar system free of remaining gas and dust.
E. Gravitational compression and radioactivity caused the earth to melt and separate into a dense iron core and a lighter mantle.

18-11. Extrasolar Planets

A. Many of the stars in the Milky Way are similar to the sun and likely have planetary systems.
B. Extrasolar planets are hard to find because:
    1. They are so far away
    2. Planets are very dim compared to stars
    3. Planets are small in size
C. More than 100 extrasolar planets have been detected..
D. Life likely exists elsewhere in the universe, although it has yet to be discovered.

18-12. Interstellar Travel

A. Interstellar travel seems impossible with existing technology because of the great distances between the stars.
B. Because of the distances involved, it is unlikely that earth has been visited by beings from another world.

18-13. Interstellar Communication

A. Existing radiotelescopes can be used to send messages to or detect radio signals from interstellar civilizations.
B. The SETI (Search for Extraterrestrial Intelligence) program uses radio telescopes to search for radio signals from extraterrestrial intelligent life.
C. So far, no contact has been made with other worlds but the hope remains.

KEY TERMS AND CONCEPTS

The questions in this section will help you review the key terms and concepts from Chapter 18.

Multiple Choice

Circle the best answer for each question.

1. The big bang that gave rise to our universe happened about
   a. 6 billion years ago
   b. 15 billion years ago
   c. 25 billion years ago
   d. 56 billion years ago

2. The Milky Way Galaxy
   a. is in the shape of a huge sphere
   b. is composed of fewer than 50 million stars
   c. is a spiral-shaped galaxy
   d. consists of stars all of the same age

3. The sun makes a complete circuit of the Milky Way once every
   a. 10,000 years
   b. 100,000 years
   c. 100 billion years
   d. 240 million years

4. The object thought to lie at the center of the Milky Way Galaxy is a
   a. pulsar
   b. Cepheid variable
   c. black hole
   d. quasar

5. The position of the sun within the Milky Way is
   a. near the galaxy's center
   b. about two-thirds of the distance from the center
   c. about one-half of the distance from the center
   d. near the galaxy's outer edge

6. Population I stars
   a. are found in the central disk of the Milky Way
   b. make up the globular clusters
   c. are all very old stars
   d. are richer in hydrogen and helium than Population II stars

7. The strongest sources of radio waves from space are
   a. pulsars
   b. galactic nebulas
   c. quasars
   d. spiral galaxies

8. Primary cosmic rays
   a. are possibly produced during supernova explosions
   b. consist mostly of neutrons
   c. are produced by radioactive substances in the earth's crust
   d. are rarely encountered by the earth

9. According to Hubble's law
   a. the more distant a galaxy, the faster it is moving away from us
   b. the more distant a galaxy, the slower it is moving away from us
   c. the closer a galaxy, the faster it is moving away from us
   d. all galaxies move toward or away from us at the same speed regardless of distance

10. Quasars
   a. have very large blue shifts
   b. are very close to the Milky Way
   c. are about the size of a galaxy
   d. show rapid fluctuations in light and/or radio outputs

11. Which of the following does not support the big bang theory of the origin of the universe?
   a. the presence of cosmic background radiation
   b. the presence of dark matter
   c. the uniform expansion of the universe
   d. the relative abundances of hydrogen and helium in the universe

12. The existence of planets outside of our own solar system is difficult to verify because of all the following reasons except
   a. planets are too dim to be seen directly
   b. planets are too small to be seen even with the best telescopes
   c. planetary systems are probably very rare throughout the universe
   d. the distances between the earth and other planetary systems are very great

True or False

Decide whether each statement is true or false. If false, either briefly state why it is false or correct the statement to make it true. See Chapter 1 or 2 for an example.

__________ 1. The galaxies are moving apart from one another, so that the universe as a whole is expanding at a steady rate.

__________ 2. Galactic nebulas are among the brightest objects in the observable universe.

__________ 3. Globular clusters are composed of Population I stars.

__________ 4. Galactic space is devoid of chemical compounds.

__________ 5. The material composition and structural pattern of the Milky Way appears to be very different from that of other galaxies.

__________ 6. Galactic spectra show that all galaxies (except a few nearby ones in the Local Group) are moving away from one another.

__________ 7. Remnants of the radiation from the early universe can be detected.

__________ 8. The protosun became a star when its internal temperature and pressure became sufficient for fusion reactions to occur.

__________ 9. A recent study of the region of the Orion Nebula by the Hubble Space Telescope indicated that about 40 percent of young stars had disks of dust and gas.

__________ 10. Interstellar travel is likely to be achieved before the end of the next century.

Fill in the Blanks

1. The concentration of stars in the disk of our galaxy appears in the night sky as the ____________________ (2 words).

2. Our galaxy is about ____________________ light-years in diameter.

3. In shape, our galaxy is classified as a ____________________ galaxy.

4. A ____________________ (2 words) is a directional antenna connected to a radio receiver that detects radio waves from space.

5. ____________________ radiation is ionizing radiation of extraterrestrial origin.

6. ____________________ (2 words) relates the red shift of a galaxy to its distance from the earth.

7. Our galaxy and about two dozen others are members of the ____________________ (2 words).

8. A massive ____________________ (2 words) is probably at the heart of every quasar.

9. A ____________________ universe eventually stops expanding and collapses because of gravity; however, an ____________________ universe continues expanding forever.

10. If the universe is closed, the universe will eventually collapse and all the matter and energy will come together in an event called the ____________________ (2 words).

11. The ____________________ (3 words) states that the universe began as a tiny fireball of dense matter that exploded billions of years ago.

12. At least 90 percent of all matter in the universe gives off no light and is known as ____________________ (2 words).

Matching

1._____ galaxy
2._____ Population I stars
3._____ globular clusters
4._____ Hubble's Law
5._____ radio telescope
6._____ cosmic rays
7._____ quasars
8._____ Population II stars
9._____ SETI
10.____ planetesimals

a. atomic nuclei that travel through the galaxy at near the speed of light
b. distant objects that are strong emitters of light and radio waves
c. NASA's search for extraterrestrial life
d. these eventually joined to form the planets
e. stars of globular clusters
f. the proportionality between galactic speed and distance
g. a directional antenna connected to a sensitive receiver
h. a rotating disk of stars
i. groups of very old stars outside the galaxy's central disk
j. stars in the central disk of our galaxy

WEB LINKS

Learn more about radio astronomy at

http://www.aoc.nrao.edu/intro/

Learn how to participate in the search for extraterrestrial intelligence at

http://setiathome.ssl.berkeley.edu/

ANSWER KEY

Multiple Choice

1. b 2. c 3. d 4. c 5. b 6. a 7. c 8. a 9. a 10. d 11. b 12. c

True or False

1. True
2. False. The brightest galactic nebula is barely visible to the naked eye; most are much fainter.
3. False. Globular clusters are composed of Population II stars.

4. False. A large number of chemical compounds have been detected in galactic space including water, carbon monoxide, ammonia, and formaldehyde among others.
5. False. The universe is uniform in material and structure and the Milky Way is no exception.
6. True
7. True
8. True
9. True
10. False. Interstellar travel seems to be unlikely in the near future and may never be possible.

Fill in the Blank

1. Milky Way
2. 130,000
3. spiral
4. radio telescope
5. cosmic
6. Hubble's law
7. Local Group
8. black hole
9. closed, open
10. big crunch
11. big bang theory
12. dark matter

Matching

1. h 2. j 3. i 4. f 5. g 6. a 7. b 8. e 9. c 10. d

SCIENCE EXPERIMENTS FOR ELEMENTARY SCHOOL TEACHERS

The following are examples of safe, easy-to-do, hands-on science experiments for the teachers of elementary school students. The only equipment and materials needed are simple tools and inexpensive household items.

Experiment 1. *Tin Can Rocket*

Punch four small holes about 1/4 inch apart near the bottom of an empty can. A soup can will do. Make sure the holes are in a straight line. Punch two more holes near the top of the can on opposite sides. Thread a wire through the two top holes and twist the two ends of the wire together. Attach a string to the wire and suspend the can over a sink or catch basin. A plastic dish pan makes a good catch basin. Fill the can with water. As the water goes out of the four bottom holes in one direction, the can swings in the opposite direction.

Pupil Investigations (may include the following processes):

1. Predicting what will happen to the can as the water is added
2. Observing the movement of the water in one direction and the movement of the can in the opposite direction.
3. Forming a theory that explains why the water moves in one direction and the can moves in the opposite direction

Experiment 1 Explained in Context (Action and Reaction Forces)

Newton's third law of motion states that when one object exerts a force on a second object, the second object exerts an equal force in the opposite direction on the first object. This law is sometimes expressed as, "for every action there is an equal and opposite reaction." As the water leaves the can in one direction, it causes the can to move in the opposite direction.

Experiment 2. *Which Is the "Warmer" Color?* Obtain two cans that are alike. Paint one can black and the other white. Fill them with equal amounts of water and place the cans in direct sunlight. Measure the temperature of the water at periodic intervals for about an hour or two.

Pupil Investigations (may include the following processes):

1. Making a graph of the water temperatures taken at 15- to 30-minute intervals for an hour or two
2. Observing that the water in the black can warms up more than the water in the white can
3. Inferring that there is a relationship between sunlight, warmth, and color

Experiment 2 Explained in Context (Radiation)

Radiation is a method of heat transmission. In that method energy travels at the speed of light from a source, through space, to an object. The best example is the sun's heat energy.

Radiation is converted to heat when it is absorbed. Thus, the air above the earth is only slightly heated by the direct sunlight, whereas all the sunlight that strikes the ground is absorbed and so heats the earth's surface. Similarly, on a cold winter day sunlight warms up the inside of a windowsill, but the glass through which the radiation travels absorbs so little radiation that it remains cold.

Of course, sunlight is not the only example of heat transmission by radiation. All objects above absolute zero actually radiate some heat, but such radiation cannot be felt until the temperatures get quite cold. Hold your hand a few inches below a clothes iron. The heat you feel is due to radiation. (Remember that if you hold your hand above the iron, the heat you feel is due to convection currents, not radiation.) Another example is to hold your hand close to a car's headlight. You should be able to feel the radiant heat.

Substances vary in their ability to absorb and reflect radiation. Black absorbs heat well but white does not. White reflects much of the energy that strikes it. As a result, the water in the black can gets warmer than the water in the white can.

Adapted from A. E. Friedl, Teaching Science to Children: An Integrated Approach, *2d ed. By permission of McGraw-Hill, Inc., New York.*

Experiment 3. *Flying Puffed Rice*

Charge a plastic spoon by vigorously rubbing it with wool cloth. Hold the charged spoon over a bowl containing puffed rice. The puffed rice will jump up and hang on the spoon and then suddenly fly away. The experiment works best when the air is cold and very dry.

Pupil Investigations (may include the following processes):

1. Observing the attraction between the puffed rice and the charged spoon
2. Discovering that the puffed rice suddenly flies away from the spoon
3. Testing to see if the experiment works if the spoon is not charged
4. Forming a theory that accounts for both the attraction and repulsion of the puffed rice

Experiment 3 Explained in Context (Static Electricity)

Rubbing the plastic spoon with wool transfers electrons from the wool to the spoon, which becomes negatively charged. The charged spoon attracts the uncharged puffed rice. Eventually, some of the electrons are transferred from the spoon into the clinging grains of puffed rice, giving them a negative charge. Since like charges repel, the puffed rice is repulsed and flies away from the spoon.

Experiment 4. *Floating Rings*. Drop rings onto a peg as shown below. Drop them one way and they pile up neatly. Drop them another way and the rings will float. Ask the children to figure out how it is done and why the rings float.

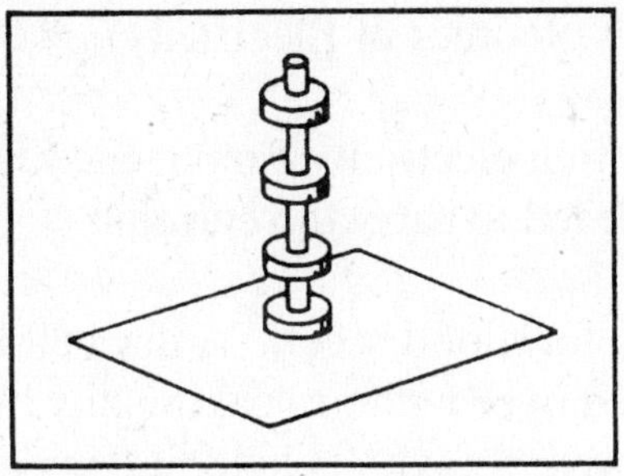

Pupil Investigation (may include the following processes):

1. Observing and feeling the rings that float on the peg
2. Inferring that the rings are actually magnets
3. Generalizing that when the rings float, it is because like poles of the magnets face and repel each other
4. Generalizing that when the rings do not float, it is because unlike poles face and attract each other

Experiment 4 Explained in Context (Attraction and Repulsion)

Floating Rings is explained by the "like poles repel" principle. The rings are really magnets with the poles on their flat surfaces. When placed over the peg with like poles facing each other, the rings repel each other and "float" in the air. If they are placed so that unlike poles face each other, the rings stick together.

Adapted from A. E. Friedl, Teaching Science to Children: An Integrated Approach, *2d ed. By permission of McGraw-Hill, Inc., New York.*

Experiment 5. *Potato Power!* Insert a copper strip and a zinc strip into a raw potato, and attach wires from the strips to a galvanometer. If a galvanometer is unavailable, touch the bare ends of the wires leading from an earphone to the copper and zinc strips. Electric current will produce static noise from the earphone. (Zinc can be obtained by cutting up the outside shell of a dry cell; copper can be obtained from a hardware store.) What does the galvanometer show? Use different materials and observe the reaction on the meter.

Pupil Investigations (may include the following processes):

1. Observing that the meter is deflected when the strips are inserted into the potato
2. Experimenting with different substances to see which will produce electrical current
3. Forming a theory about the source of the current

Experiment 5 Explained in Context (Sources of Electricity)

Potato Power! is an example of getting electricity from a chemical source. The potato contains dissimilar metals and the juices needed to carry the current.

In a sense the potato cell is really a form of dry cell. A dry cell is not really dry inside. It is called "dry" because it can be tipped over without leaking any liquids. The potato cell can work in the same way. Regardless of how the potato is held, it does not spill.

In the place of a potato you may use a lemon. In fact, a lemon works better in some ways. The lemon juice is a better conductor, so it produces a somewhat stronger current. Unfortunately, the juices tend to run and drip.

Adapted from A. E. Friedl, Teaching Science to Children: An Integrated Approach, *2d ed. By permission of McGraw-Hill, Inc., New York.*

Experiment 6. *Make the Lion Roar*. Punch a tiny hole in the bottom of an empty milk carton and insert a string. Fasten the string to a button so that it will not pull out of the hole. Hold the string tightly and stroke it with a damp cloth or paper towel. What does it sound like? (You may try the same activity with a tin can to make "the rooster crow.")

Pupil Investigations (may include the following processes):

1. Experimenting with the device to determine the kinds of sounds that can be created
2. Observing that the sound caused by rubbing is very loud
3. Inferring a relationship between the size of the vibrating surface and the loudness of sound

Experiment 6 Explained in Context (Volume or Loudness)

Make the Lion Roar is an example of forced vibration. The string has a very small surface. So, when you rub it with a damp cloth or paper, it gives off a feeble, squeaky sound. When the sound goes into the milk carton, however, a much larger surface is set in motion, and the sound is much louder. The carton will produce a lion's roar. If you use a tin can, it will produce the sound of a crowing rooster.

Adapted from A. E. Friedl, Teaching Science to Children: An Integrated Approach, *2d ed. By permission of McGraw-Hill, Inc., New York.*

Experiment 7. *Ghostly Candle.* Set a mirror on a sheet of paper as shown below. Draw a line at right angles to the mirror, and place a candle to the left and a few centimeters in front of the mirror. Go to the right-hand side of the paper, and look just barely over the surface of the paper at the image of the candle.

Place a second candle behind the mirror to show where the image of the first candle appears to be. Measure the angles and distances from the mirror to the candle. Compare the angles and distances of the first candle with the angles and distances of the second candle.

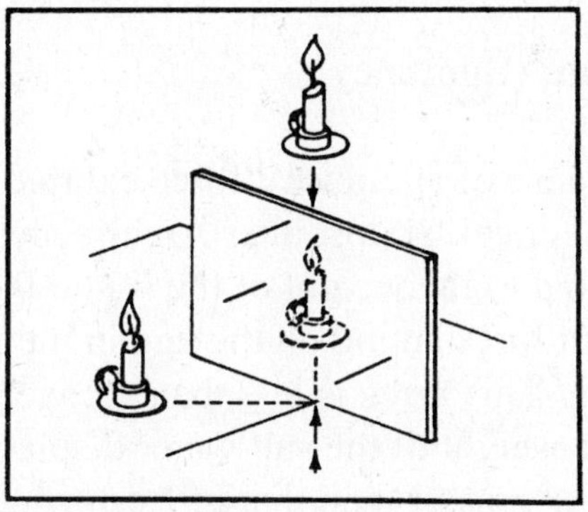

Pupil Investigations (may include the following processes):

1. Measuring the distance between the first candle and mirror
2. Measuring the distance between the second candle and mirror
3. Measuring the angles involved with both candles
4. Making a rule about the reflection of light

Experiment 7 Explained in Context (Light Can Be Reflected)

The angles of incidence and reflection are shown in Ghostly Candle. If the lines are drawn carefully, the distances and angles should be identical. Another trait of even reflection is that images are "folded over" from left to right. That is, if you stand in front of a plane mirror and wink your right eye, your image will wink its left eye. The same folding over characteristic can be seen in reflections of printed words. However, images are not folded over top to bottom.

Adapted from A. E. Friedl, Teaching Science to Children: An Integrated Approach, *2d ed. By permission of McGraw-Hill, Inc., New York.*

Experiment 8. *The Levitating Egg*

Obtain two identical wide-mouth jars. Half-fill one jar with tap water. Half-fill the other with salt water (about six tablespoons of salt to a pint of tap water). Place a fresh egg in the first jar; it will sink to the bottom. Now add the water from the second jar to the first by carefully pouring it over a spoon so that the two liquids do not mix. Mysteriously, the egg will float as if levitated. Ask the children to explain the results.

Pupil Investigations (may include the following processes):

1. Observing that the egg at first sinks to the bottom of the jar and then floats when the salt water is added
2. Inferring that adding the salt water caused the egg to float
3. Testing to see if an egg will sink or float in salt water
4. Forming a theory to explain why the egg was "levitated" when the salt water was added

Experiment 8 Explained in Context (Buoyancy)

An object that floats does so because of an upward force exerted by the liquid. This force is known as buoyancy. The ancient Greek philosopher Archimedes discovered that the buoyant force acting upon an object is equal to the weight of the liquid that is displaced by the object. In the Levitating Egg, the buoyant force acting on the egg in the fresh water is equal to the weight of the water displaced by the egg. This force is less than the weight of the egg, so the egg sinks. When the salt water is added, the weight of the salt water displaced is greater than the weight of the egg (the density of salt water is greater than that of fresh water) and the egg floats at the salt water/fresh water boundary.

Experiment 9. *Soap-Powered Boat*

Make a boat by cutting a triangle about 1-in high from a piece of shirt cardboard. Cut a small notch in the center of the base of the triangle. Fill a plastic dishpan with water and place the boat at one end. Place a drop of liquid dishwashing detergent in the notch at the base of the triangle. The boat will move across the water's surface.

Pupil Investigations (may include the following):

1. Observing that the boat does not move until the soap is added
2. Experimenting with other substances (water, salad oil, grape juice) to see if they "power" the boat
3. Forming a theory to explain how the soap propelled the boat

Experiment 9 Explained in Context (Cohesion)

Cohesion is the force of attraction between molecules of the same substance. Liquids, such as water, exhibit a special form of cohesion at their surfaces known as surface tension. The soap breaks the surface tension of the water, causing a backward movement of the water and a forward movement of the boat.

Experiment 10. *Direct and Slant Rays*. Draw parallel lines on a sheet of cardboard to represent rays from the sun. Cut out an arc on the cardboard that corresponds to the curvature of a globe. Hold the curved end of the cardboard against the globe so that one set of five lines (sun's rays) is on the equator and the other set of lines is at the north pole.

Note how much area is covered by the five rays at the equator, and compare with the rays at the pole. How does this show why polar regions are cold and the equator warm?

Pupil Investigations (may include the following processes):

1. Measuring the area covered by the five rays at the equator and at the north pole
2. Discovering that the solar rays at the equator are high in the sky
3. Discovering that the solar rays at the north pole are low in the sky
4. Generalizing that the sun's energy is high at the equator because the rays are concentrated but low at the poles because the rays are spread out

Experiment 10 Explained in Context (The Seasons)

Sunlight strikes the polar region at an angle. Solar energy is spread out over a larger area at the poles than at the equator.

Pick four or five lines and see how much area is covered at the equator. Then measure the areas at the pole covered by the same number of lines. You will find that the rays are spread out and diluted at the poles, keeping the area from getting warm.

One final factor makes the problem even worse. The sunlight is not only diluted at the poles, but it is also filtered and weakened by the many hundreds of kilometers of atmosphere that it must pass through on its slanted path to the earth. At the equator the sun takes a much shorter and more direct path to the earth.

Adapted from A. E. Friedl, Teaching Science to Children: An Integrated Approach, *2d ed. By permission of McGraw-Hill, Inc., New York.*

WEB LINKS

The Science Spot: Physics Lesson Plans Links

http://sciencespot.net/Pages/classphyslsn.html

Science Education Outreach: Physics Demonstrations, Lectures, and Workshops

http://www.lis.uiuc.edu/~chip/pubs/97physics.shtml

Science Activities, Labs, and Lesson Plans

http://sciencepage.org/lessons.htm

Elementary Science Education Resources on the Internet

http://www.uvm.edu/~jmorris/Sci.html

Internet-Based Lesson Plans

http://www.col-ed.org/smcnws/msres/curriculum.html

Teaching Earth Science

http://www.earth.nasa.gov/education/

ProTeacher Science Activities and Lesson Plans

http://www.proteacher.com/110000.shtml

Earthquake Resources for Kindergarten and Elementary School

http://quake.ualr.edu/schools/elem.htm

Elementary Teacher's Resource Sites

http://www.ket.org/Education/IN/elem.html

Notes

Notes

Notes

Notes

Notes